ISEE Middle-Level

Subject Test Mathematics

Student Practice Workbook

+ Two Full-Length ISEE Middle-Level Math Tests

Math Notion

www.MathNotion.com

ISEE Middle-Level Subject Test Mathematics

ISEE Middle-Level Subject Test Mathematics

ISEE Middle-Level Subject Test Mathematics

Published in the United State of America By

The Math Notion

Web: WWW.MathNotion.com

Email: info@Mathnotion.com

Copyright © 2021 by the Math Notion. All rights reserved. No part of this publication may be reproduced, stored in a retrieval system, or transmitted in any form or by any means, electronic, mechanical, photocopying, recording, scanning, or otherwise, except as permitted under Section 107 or 108 of the 1976 United States Copyright Ac, without permission of the author.

All inquiries should be addressed to the Math Notion.

ISBN: 978-1-63620-080-4

The Math Notion

Michael Smith has been a math instructor for over a decade now. He launched the Math Notion. Since 2006, we have devoted our time to both teaching and developing exceptional math learning materials. As a test prep company, we have worked with thousands of students. We have used the feedback of our students to develop a unique study program that can be used by students to drastically improve their math scores fast and effectively. We have more than a thousand Math learning books including:

- SAT Math Prep
- ACT Math Prep
- SSAT/ISEE Math Prep
- Accuplacer Math Prep
- Common Core Math Prep
- many Math Education Workbooks, Study Guides, Practice and Exercise Books

As an experienced Math test preparation company, we have helped many students raise their standardized test scores—and attend the colleges of their dreams: We tutor online and in person, we teach students in large groups, and we provide training materials and textbooks through our website and through Amazon.

You can contact us via email at:

info@Mathnotion.com

ISEE Middle-Level Subject Test Mathematics

Get the Targeted Practice You Need to Ace the ISEE Middle-Level Math Test!

ISEE Middle-Level Subject Test Mathematics includes easy-to-follow instructions, helpful examples, and plenty of math practice problems to assist students to master each concept, brush up their problem-solving skills, and create confidence. The ISEE Middle-Level math practice book provides numerous opportunities to evaluate basic skills along with abundant remediation and intervention activities. It is a skill that permits you to quickly master intricate information and produce better leads in less time.

Students can boost their test-taking skills by taking the book's two practice ISEE Middle-Level Math exams. All test questions answered and explained in detail.

Important Features of the ISEE Middle-Level Math Book:

- A **complete review** of ISEE Middle-Level math test topics,
- Over 2,500 practice problems covering all topics tested,
- The most important concepts you need to know,
- Clear and concise, easy-to-follow sections,
- Well designed for enhanced learning and interest,
- Hands-on experience with all question types,
- **2 full-length practice tests** with detailed answer explanations,
- Cost-Effective Pricing,

Powerful math exercises to help you avoid traps and pacing yourself to beat the ISEE Middle-Level test. Students will gain valuable experience and raise their confidence by taking ISEE math practice tests, learning about test structure, and gaining a deeper understanding of what is tested on the ISEE Middle-Level math. If ever there was a book to respond to the pressure to increase students' test scores, this is it.

ISEE Middle-Level Subject Test Mathematics

WWW.MathNotion.COM

… So Much More Online!

- ✓ FREE Math Lessons
- ✓ More Math Learning Books!
- ✓ Mathematics Worksheets
- ✓ Online Math Tutors

For a PDF Version of This Book

Please Visit WWW.MathNotion.com

ISEE Middle-Level Subject Test Mathematics

Contents

Chapter 1 : Review of the Whole Number Operations 11
- Adding Whole Numbers 12
- Subtracting Whole Numbers 13
- Multiplying Whole Numbers..14
- Dividing Hundreds..15
- Long Division by Two Digits 16
- Division with Remainders 16
- Rounding Whole Numbers 17
- Whole Number Estimation 18
- Answers of Worksheets 19

Chapter 2 : Integers and Number Theory 21
- Adding and Subtracting Integers 22
- Multiplying and Dividing Integers 23
- Order of Operations 24
- Ordering Integers and Numbers...25
- Integers and Absolute Value 26
- Factoring Numbers 27
- Prime Factorization...27
- Divisibility Rules..28
- Greatest Common Factor 29
- Least Common Multiple 30
- Answers of Worksheets 31

Chapter 3 : Fractions 35
- Simplifying Fractions...36
- Adding and Subtracting Fractions 37
- Multiplying and Dividing Fractions 38
- Adding and Subtracting Mixed Numbers 39
- Multiplying and Dividing Mixed Numbers 40
- Answers of Worksheets 41

Chapter 4 : Decimals 43
- Adding and Subtracting Decimals..44
- Multiplying and Dividing Decimals 45
- Comparing Decimals..46

WWW.MathNotion.Com

ISEE Middle-Level Subject Test Mathematics

    Rounding Decimals 47
    Convert Fraction to Decimal 48
    Convert Decimal to Percent 49
    Convert Fraction to Percent 50
    Answers of Worksheets 51

Chapter 5 : Proportions, Ratios, and Percent 53
    Simplifying Ratios 54
    Proportional Ratios 55
    Similarity and Ratios 56
    Ratio and Rates Word Problems 57
    Percentage Calculations 58
    Percent Problems 59
    Discount, Tax and Tip 60
    Answers of Worksheets 61

Chapter 6 : Exponents and Radicals Expressions 63
    Adding and Subtracting Exponents 64
    Multiplication Property of Exponents 65
    Zero and Negative Exponents 66
    Division Property of Exponents 67
    Powers of Products and Quotients 68
    Negative Exponents and Negative Bases 69
    Scientific Notation 70
    Square Roots 71
    Answers of Worksheets 72

Chapter 7 : Measurements 75
    Reference Measurement 76
    Metric Length Measurement 77
    Customary Length Measurement 77
    Metric Capacity Measurement 78
    Customary Capacity Measurement 78
    Metric Weight and Mass Measurement 79
    Customary Weight and Mass Measurement 79
    Temperature 80
    Time 81
    Answers of Worksheets 82

Chapter 8 : Algebraic Expressions 84
    Find a Rule! 85

WWW.MathNotion.Com

ISEE Middle-Level Subject Test Mathematics

Translate Phrases into an Algebraic Statement 86
Simplifying Variable Expressions 87
The Distributive Property 88
Evaluating One Variable Expressions 89
Combining like Terms 90
Answers of Worksheets 91

Chapter 9 : Equations and Inequalities 93
One–Step Equations 94
One–Step Equation Word Problems 95
Two-Steps Equations 96
Multi–Step Equations 97
One-Step Inequalities 98
Graphing Inequalities 99
Two-Steps Inequality 100
Multi-Step Inequalities 101
Answers of Worksheets 102

Chapter 10 : Geometry and Solid Figures 105
Angles 106
Pythagorean Relationship 107
Triangles 108
Polygons 109
Trapezoids 110
Circles 111
Cubes 112
Rectangular Prism 113
Cylinder 114
Answers of Worksheets 115

Chapter 11 : Statistics and Probability 117
Mean and Median 118
Mode and Range 119
Times Series 120
Stem–and–Leaf Plot 121
Quartile of a Data Set 122
Box and Whisker Plots 122
Pie Graph 123
Probability Problems 124
Answers of Worksheets 125

WWW.MathNotion.Com

ISEE Middle-Level Subject Test Mathematics

Chapter 12 : ISEE Middle Level Practice Tests **127**
ISEE Middle Level Practice Test Answer Sheets 129
ISEE Middle Level Practice Test 1 131
Quantitative Reasoning 131
Mathematics Achievement 140
ISEE Middle Level Practice Test 2 151
Quantitative Reasoning 151
Mathematics Achievement 161

Chapter 13 : Answers and Explanations **173**
Answer Key 173
Score Your Test 175
Practice Tests 1: Quantitative Reasoning 177
Practice Tests 1: Mathematics Achievement 183
Practice Tests 2: Quantitative Reasoning 189
Practice Tests 2: Mathematics Achievement 195

ISEE Middle-Level Subject Test Mathematics

Chapter 1 : Review of the Whole Number Operations

Topics that you'll learn in this chapter:

- ✓ Adding Whole Numbers
- ✓ Subtracting Whole Numbers
- ✓ Multiplying Whole Numbers
- ✓ Dividing Hundreds
- ✓ Long Division by One Digit
- ✓ Division with Remainders
- ✓ Rounding Whole Numbers
- ✓ Whole Number Estimation

"Wherever there is number, there is beauty." –Proclus

ISEE Middle-Level Subject Test Mathematics

Adding Whole Numbers

✍ Add.

1) 5,763
   + 8,238
   _____

2) 6,834
   + 4,998
   _____

3) 3,548
   + 5,693
   _____

4) 2,769
   +8,872
   _____

5) 3,196
   +2,936
   _____

6) 7,009
   + 4,992
   _____

✍ Find the missing numbers.

7) 3,468 + ___ = 4,102

8) 840 + 2,360 = ___

9) 5,200 + ___ = 7,980

10) 631 + ___ = 2,007

11) ___ + 803 = 3,945

12) ___ + 2,156 = 5,922

13) David sells gems. He finds a diamond in Istanbul and buys it for $4,795. Then, he flies to Cairo and purchases a bigger diamond for the bargain price of $9,633. How much does David spend on the two diamonds? _____

WWW.MathNotion.Com

ISEE Middle-Level Subject Test Mathematics

Subtracting Whole Numbers

✎ Subtract.

1) 10,512
   − 4,411
   _____

2) 5,204
   − 3,679
   _____

3) 8,520
   − 6,483
   _____

4) 8,001
   − 5,224
   _____

5) 11,916
   − 8,711
   _____

6) 5,005
   − 2,008
   _____

✎ Find the missing number.

7) 5,263 − ___ = 2,367

8) 7,198 − ___ = 4,742

9) 8,928 − 3,764 = ___

10) 6,511 − ___ = 3,759

11) 7,003 − 5,489 = ___

12) 8,800 − 5,995 = ___

13) Jackson had $7,189 invested in the stock market until he lost $3,793 on those investments. How much money does he have in the stock market now?

_____

WWW.MathNotion.Com

Multiplying Whole Numbers

✎ **Find the answers.**

1) 2,200 × 31

2) 3,200 × 22

3) 5,790 × 5

4) 5,220 × 3

5) 6,911 × 3

6) 1,998 × 40

7) 2,893 × 5.5

8) 2,254 × 3.5

9) 4,372 × 4.8

10) 3,984 × 2.75

11) 4,900 × 2.5

12) 8,200 × 4.5

Dividing Hundreds

✍ **Find answers.**

1) $4,440 \div 400$

2) $1,600 \div 40$

3) $9,990 \div 90$

4) $4,200 \div 60$

5) $6,400 \div 8,000$

6) $2,700 \div 30$

7) $3,333 \div 30$

8) $558 \div 45$

9) $2,278 \div 85$

10) $1,683 \div 55$

11) $1,582 \div 35$

12) $9,000 \div 600$

13) $1,000 \div 2,500$

14) $44.8 \div 20$

15) $6,800 \div 400$

16) $1,500 \div 5,000$

17) $36.60 \div 120$

18) $7,700 \div 700$

19) $5,400 \div 600$

20) $8,000 \div 160$

21) $18,000 \div 9,000$

22) $42,000 \div 30$

23) $480 \div 40$

24) $63,000 \div 900$

ISEE Middle-Level Subject Test Mathematics

Long Division by Two Digits

✏️ Find the quotient.

1) 18)576   10) 41)1,476

2) 14)952   11) 53)2,491

3) 21)588   12) 60)2,880

4) 23)299   13) 32)2,912

5) 44)748   14) 77)8,393

6) 26)234   15) 85)3,740

7) 16)496   16) 57)4,617

8) 29)1,479   17) 50)9,200

9) 54)1,080   18) 25)15,400

Division with Remainders

✏️ Find the quotient with remainder.

1) 14)715   8) 65)8,624

2) 16)2,750   9) 35)5,705

3) 27)4,603   10) 92)13,161

4) 58)2,554   11) 46)12,214

5) 42)7,732   12) 69)42,482

6) 63)6,737   13) 85)6,858

7) 71)9,036   14) 87)34,304

WWW.MathNotion.Com

ISEE Middle-Level Subject Test Mathematics

Rounding Whole Numbers

✎ Round each number to the underlined place value.

1) 7,5̲33

2) 9,3̲74

3) 8,8̲3

4) 2,3̲6̲8

5) 5,5̲7̲7

6) 3,3̲8̲1

7) 3,5̲20

8) 9,3̲3̲8

9) 8.5̲81

10) 33.5̲7

11) 51.6̲9

12) 22.1̲38

13) 6̲,758

14) 11,5̲7

15) 8,8̲3̲8

16) 5.8̲89

17) 1.8̲60

18) 25.0̲70

19) 9̲.332

20) 49.4̲8

21) 28.8̲9

22) 24,3̲7̲7

23) 52,1̲5̲8

24) 13,8̲8̲3

25) 9,̲609

26) 17,4̲5̲1

27) 18,7̲68

ISEE Middle-Level Subject Test Mathematics

Whole Number Estimation

✎ **Estimate the sum by rounding each added to the nearest ten.**

1) 875 + 325

2) 985 + 1,452

3) 2,424 + 4,128

4) 1,576 + 6,279

5) 1,247 + 3,863

6) 6,746 + 5,121

7) 3,924 + 6,456

8) 1,785 + 7,164

9) $\begin{array}{r} 1,458 \\ +\ 2,442 \\ \hline \end{array}$

10) $\begin{array}{r} 5,689 \\ +\ 4,151 \\ \hline \end{array}$

11) $\begin{array}{r} 8,259 \\ +\ 4,754 \\ \hline \end{array}$

12) $\begin{array}{r} 6,788 \\ +\ 3,954 \\ \hline \end{array}$

13) $\begin{array}{r} 9,123 \\ +\ 4,455 \\ \hline \end{array}$

14) $\begin{array}{r} 6,680 \\ +\ 5,358 \\ \hline \end{array}$

15) $\begin{array}{r} 3,165 \\ +\ 7,124 \\ \hline \end{array}$

16) $\begin{array}{r} 8,859 \\ +\ 6,452 \\ \hline \end{array}$

WWW.MathNotion.Com

ISEE Middle-Level Subject Test Mathematics

Answers of Worksheets

Adding Whole Numbers

1) 14,001
2) 11,832
3) 9,241
4) 11,641
5) 6,132
6) 12,001
7) 634
8) 3,200
9) 2,780
10) 1,376
11) 3,142
12) 3,766
13) $14,428

Subtracting Whole Numbers

1) 6,101
2) 1,525
3) 2,037
4) 2,777
5) 3,205
6) 2,997
7) 2,896
8) 2,456
9) 5,164
10) 2,752
11) 1,514
12) 2,805
13) 3,396

Multiplying Whole Numbers

1) 68,200
2) 70,400
3) 28,950
4) 15,660
5) 20,733
6) 79,920
7) 15,911.5
8) 7,889
9) 20,985.6
10) 10,956
11) 12,250
12) 36,900

Dividing Hundreds

1) 11.1
2) 40
3) 111
4) 70
5) 0.8
6) 90
7) 111.1
8) 12.4
9) 26.8
10) 30.6
11) 45.2
12) 15
13) 0.4
14) 2.24
15) 17
16) 0.3
17) 0.305
18) 11
19) 9
20) 50
21) 2
22) 1,400
23) 12
24) 70

Long Division by Two Digits

1) 32
2) 68
3) 28
4) 13
5) 17
6) 9
7) 31
8) 51
9) 20
10) 36
11) 47
12) 48
13) 91
14) 109
15) 44
16) 81
17) 184
18) 616

WWW.MathNotion.Com

ISEE Middle-Level Subject Test Mathematics

Division with Remainders

1) 51 R1
2) 171 R14
3) 170 R13
4) 44 R2
5) 184 R4
6) 106 R59
7) 127 R19
8) 132 R44
9) 163 R0
10) 143 R5
11) 265 R24
12) 615 R47
13) 80 R58
14) 394 R26

Rounding Whole Numbers

1) 7,500
2) 9,400
3) 8,880
4) 2,370
5) 5,580
6) 3,380
7) 3,500
8) 9,340
9) 8.60
10) 33.60
11) 51.70
12) 22.100
13) 7,000
14) 11,560
15) 8,840
16) 5.900
17) 1.900
18) 25.100
19) 9.000
20) 49.50
21) 28.90
22) 24,380
23) 52,160
24) 13,880
25) 9,600
26) 17,450
27) 18,800

Whole Number Estimation

1) 1,200
2) 2,440
3) 6,550
4) 7,860
5) 5,110
6) 11,870
7) 10,380
8) 8,950
9) 3,900
10) 9,840
11) 13,010
12) 10,740
13) 13,580
14) 12,040
15) 10,290
16) 15,310

ISEE Middle-Level Subject Test Mathematics

Chapter 2:
Integers and Number Theory

Topics that you will practice in this chapter:

- ✓ Adding and Subtracting Integers
- ✓ Multiplying and Dividing Integers
- ✓ Order of Operations
- ✓ Ordering Integers and Numbers
- ✓ Integers and Absolute Value
- ✓ Factoring Numbers
- ✓ Prime Factorization
- ✓ Divisibility Rules
- ✓ Greatest Common Factor (GCF)
- ✓ Least Common Multiple (LCM)

"In order to gain the most, you have to know how to convert Negatives to Positives."

–Stubborn Clown

ISEE Middle-Level Subject Test Mathematics

Adding and Subtracting Integers

✎ **Find each sum.**

1) $14 + (-6) =$

2) $(-13) + (-20) =$

3) $5 + (-28) =$

4) $50 + (-12) =$

5) $(-7) + (-15) + 3 =$

6) $30 + (-14) + 8 =$

7) $40 + (-10) + (-14) + 17 =$

8) $(-15) + (-20) + 13 + 35 =$

9) $40 + (-20) + (38 - 29) =$

10) $28 + (-12) + (30 - 12) =$

✎ **Find each difference.**

11) $(-18) - (-7) =$

12) $25 - (-14) =$

13) $(-20) - 36 =$

14) $34 - (-19) =$

15) $51 - (30 - 21) =$

16) $17 - (5) - (-24) =$

17) $(35 + 20) - (-46) =$

18) $48 - 16 - (-8) =$

19) $62 - (28 + 17) - (-15) =$

20) $58 - (-23) - (-31) =$

21) $19 - (-8) - (-13) =$

22) $(19 - 24) - (-14) =$

23) $27 - 33 - (-21) =$

24) $58 - (32 + 24) - (-9) =$

25) $36 - (-30) + (-17) =$

26) $27 - (-42) + (-31) =$

WWW.MathNotion.Com

ISEE Middle-Level Subject Test Mathematics

Multiplying and Dividing Integers

✎ **Find each product.**

1) $(-9) \times (-5) =$

2) $(-3) \times 9 =$

3) $8 \times (-12) =$

4) $(-7) \times (-20) =$

5) $(-3) \times (-5) \times 6 =$

6) $(14 - 3) \times (-8) =$

7) $12 \times (-9) \times (-3) =$

8) $(140 + 10) \times (-2) =$

9) $10 \times (-12 + 8) \times 3 =$

10) $(-8) \times (-5) \times (-10) =$

✎ **Find each quotient.**

11) $42 \div (-7) =$

12) $(-48) \div (-6) =$

13) $(-40) \div (-8) =$

14) $54 \div (-2) =$

15) $152 \div 19 =$

16) $(-144) \div (-12) =$

17) $180 \div (-10) =$

18) $(-312) \div (-12) =$

19) $221 \div (-13) =$

20) $(-126) \div (6) =$

21) $(-161) \div (-7) =$

22) $-266 \div (-14) =$

23) $(-120) \div (-4) =$

24) $270 \div (-18) =$

25) $(-208) \div (-8) =$

26) $(135) \div (-15) =$

ISEE Middle-Level Subject Test Mathematics

Order of Operations

✏ **Evaluate each expression.**

1) $7 + (5 \times 4) =$

2) $14 - (3 \times 6) =$

3) $(19 \times 4) + 16 =$

4) $(16 - 7) - (8 \times 2) =$

5) $27 + (18 \div 3) =$

6) $(18 \times 8) \div 6 =$

7) $(32 \div 4) \times (-2) =$

8) $(9 \times 4) + (32 - 18) =$

9) $24 + (4 \times 3) + 7 =$

10) $(36 \times 3) \div (2 + 2) =$

11) $(-7) + (12 \times 3) + 11 =$

12) $(8 \times 5) - (24 \div 6) =$

13) $(7 \times 6 \div 3) - (12 + 9) =$

14) $(13 + 5 - 14) \times 3 - 2 =$

15) $(20 - 14 + 30) \times (64 \div 4) =$

16) $32 + (28 - (36 \div 9)) =$

17) $(7 + 6 - 4 - 7) + (15 \div 5) =$

18) $(85 - 20) + (20 - 18 + 7) =$

19) $(20 \times 2) + (14 \times 3) - 22 =$

20) $18 + 5 - (30 \times 3) + 20 =$

21) $(\frac{7}{5 - 1}) \times (2 + 6) \times 2$

22) $20 \div (4 - (10 - 8))$

WWW.MathNotion.Com

ISEE Middle-Level Subject Test Mathematics

Ordering Integers and Numbers

✎ **Order each set of integers from least to greatest.**

1) $8, -10, -5, -3, 4$ ___, ___, ___, ___, ___, ___

2) $-10, -18, 6, 14, 27$ ___, ___, ___, ___, ___, ___

3) $15, -8, -21, 21, -23$ ___, ___, ___, ___, ___, ___

4) $-14, -40, 23, -12, 47$ ___, ___, ___, ___, ___, ___

5) $59, -54, 32, -57, 36$ ___, ___, ___, ___, ___, ___

6) $68, 26, -19, 47, -34$ ___, ___, ___, ___, ___, ___

✎ **Order each set of integers from greatest to least.**

7) $18, 36, -16, -18, -10$ ___, ___, ___, ___, ___, ___

8) $27, 34, -12, -24, 94$ ___, ___, ___, ___, ___, ___

9) $50, -21, -13, 42, -2$ ___, ___, ___, ___, ___, ___

10) $37, 46, -20, -16, 86$ ___, ___, ___, ___, ___, ___

11) $-18, 88, -26, -59, 75$ ___, ___, ___, ___, ___, ___

12) $-65, -30, -25, 3, 14$ ___, ___, ___, ___, ___, ___

WWW.MathNotion.Com

ISEE Middle-Level Subject Test Mathematics

Integers and Absolute Value

✎ **Write absolute value of each number.**

1) $|-2| =$

2) $|-27| =$

3) $|-20| =$

4) $|14| =$

5) $|6| =$

6) $|-55| =$

7) $|16| =$

8) $|2| =$

9) $|54| =$

10) $|-4| =$

11) $|-11|$

12) $|88| =$

13) $|0| =$

14) $|79| =$

15) $|-32| =$

16) $|-17| =$

17) $|42| =$

18) $|-46| =$

19) $|1| =$

20) $|-40| =$

✎ **Evaluate the value.**

21) $|-5| - \frac{|-21|}{7} =$

22) $14 - |3 - 15| - |-4| =$

23) $\frac{|-32|}{4} \times |-4| =$

24) $\frac{|7 \times (-3)|}{7} \times \frac{|-19|}{3} =$

25) $|4 \times (-5)| + \frac{|-40|}{5} =$

26) $\frac{|-45|}{9} \times \frac{|-24|}{12} =$

27) $|-12 + 8| \times \frac{|-7 \times 7|}{7} =$

28) $\frac{|-11 \times 2|}{4} \times |-16| =$

WWW.MathNotion.Com

ISEE Middle-Level Subject Test Mathematics

Factoring Numbers

✎ List all positive factors of each number.

1) 12	6) 56	11) 27
2) 16	7) 65	12) 63
3) 28	8) 70	13) 72
4) 34	9) 25	14) 15
5) 95	10) 48	15) 80

✎ List the prime factorization for each number.

16) 10	19) 30	22) 55
17) 26	20) 40	23) 78
18) 20	21) 44	24) 96

Prime Factorization

✎ Factor the following numbers to their prime factors.

1) 6	9) 58	17) 69
2) 49	10) 62	18) 76
3) 60	11) 75	19) 86
4) 4	12) 88	20) 92
5) 46	13) 93	21) 99
6) 57	14) 100	22) 77
7) 54	15) 68	23) 90
8) 38	16) 90	24) 74

WWW.MathNotion.Com

ISEE Middle-Level Subject Test Mathematics

Divisibility Rules

✎ Use the divisibility rules to underline the factors of the number.

1) 8   2 3 4 5 6 7 8 9 10

2) 18   2 3 4 5 6 7 8 9 10

3) 55   2 3 4 5 6 7 8 9 10

4) 45   2 3 4 5 6 7 8 9 10

5) 20   2 3 4 5 6 7 8 9 10

6) 9   2 3 4 5 6 7 8 9 10

7) 21   2 3 4 5 6 7 8 9 10

8) 28   2 3 4 5 6 7 8 9 10

9) 36   2 3 4 5 6 7 8 9 10

10) 40   2 3 4 5 6 7 8 9 10

11) 39   2 3 4 5 6 7 8 9 10

12) 51   2 3 4 5 6 7 8 9 10

WWW.MathNotion.Com

ISEE Middle-Level Subject Test Mathematics

Greatest Common Factor

✎ Find the GCF for each number pair.

1) 6, 2

2) 4, 5

3) 3, 12

4) 7, 3

5) 5, 10

6) 8, 48

7) 6, 18

8) 9, 15

9) 12, 18

10) 4, 36

11) 6, 10

12) 28, 52

13) 25, 10

14) 22, 24

15) 9, 54

16) 8, 54

17) 42, 14

18) 16, 40

19) 9, 2, 3

20) 5, 15, 10

21) 7, 9, 2

22) 16, 64

23) 30, 48

24) 36, 63

Least Common Multiple

✏ **Find the LCM for each number pair.**

1) 6, 9

2) 15, 45

3) 16, 40

4) 12, 36

5) 18, 27

6) 14, 42

7) 6, 30

8) 8, 56

9) 7, 21

10) 8, 20

11) 15, 25

12) 7, 9

13) 4, 11

14) 8, 28

15) 28, 56

16) 40, 50

17) 12, 13

18) 22, 11

19) 36, 20

20) 15, 35

21) 18, 81

22) 30, 54

23) 18, 45

24) 75, 25

ISEE Middle-Level Subject Test Mathematics

Answers of Worksheets

Adding and Subtracting Integers

1) 8
2) −33
3) −23
4) 38
5) −19
6) 24
7) 33
8) 13
9) 29
10) 34
11) −11
12) 39
13) −56
14) 53
15) 42
16) 36
17) 101
18) 40
19) 32
20) 112
21) 40
22) 9
23) 15
24) 11
25) 49
26) 38

Multiplying and Dividing Integers

1) 45
2) −27
3) −96
4) 140
5) 90
6) −88
7) 324
8) −300
9) −120
10) −400
11) −6
12) 8
13) 5
14) −27
15) 8
16) 12
17) −18
18) 26
19) −17
20) −21
21) 23
22) 19
23) 30
24) −15
25) 26
26) −9

Order of Operations

1) 27
2) −4
3) 92
4) −7
5) 33
6) 24
7) −16
8) 50
9) 43
10) 27
11) 40
12) 36
13) −7
14) 10
15) 576
16) 56
17) 5
18) 74
19) 60
20) −47
21) 28
22) 10

Ordering Integers and Numbers

1) −10, −5, −3, 4, 8
2) −18, −10, 6, 14, 27
3) −23, −21, −8, 15, 21
4) −40, −14, −12, 23, 47
5) −57, −54, 32, 36, 59
6) −34, −19, 26, 47, 68
7) 36, 18, −10, −16, −18
8) 94, 34, 27, −12, −24
9) 50, 42, −2, −13, −21
10) 86, 46, 37, −16, −20
11) 88, 75, −18, −26, −59
12) 14, 3, −25, −30, −65

WWW.MathNotion.Com

ISEE Middle-Level Subject Test Mathematics

Integers and Absolute Value

1) 2	8) 2	15) 32	22) −2
2) 27	9) 54	16) 17	23) 32
3) 20	10) 4	17) 42	24) 19
4) 14	11) 11	18) 46	25) 28
5) 6	12) 88	19) 1	26) 10
6) 55	13) 0	20) 40	27) 28
7) 16	14) 79	21) 2	28) 88

Factoring Numbers

1) 1, 2, 3, 4, 6, 12
2) 1, 2, 4, 8, 16
3) 1, 2, 4, 7, 14, 28
4) 1, 2, 17, 34
5) 1, 5, 19, 95
6) 1, 2, 4, 7, 8, 14, 28, 56
7) 1, 5, 13, 65
8) 1, 2, 5, 7, 10, 14, 35, 70
9) 1, 5, 25
10) 1, 2, 3, 4, 6, 8, 12, 16, 24, 48
11) 1, 3, 9, 27
12) 1, 3, 7, 9, 21, 63
13) 1, 2, 3, 4, 6, 8, 9, 12, 18, 24, 36, 72
14) 1, 3, 5, 15
15) 1, 2, 4, 5, 8, 10, 16, 20, 40, 80
16) 2×5
17) 2×13
18) $2 \times 2 \times 5$
19) $2 \times 3 \times 5$
20) $2 \times 2 \times 2 \times 5$
21) $2 \times 2 \times 11$
22) 5×11
23) $2 \times 3 \times 13$
24) $2 \times 2 \times 2 \times 2 \times 2 \times 3$

Prime Factorization

1) 2. 3	9) 2. 29	17) 3. 23
2) 7. 7	10) 2. 31	18) 2. 2. 19
3) 2. 2. 3. 5	11) 3. 5. 5	19) 2. 43
4) 2. 2	12) 2. 2. 2. 11	20) 2. 2. 23
5) 2. 23	13) 3. 31	21) 3. 3. 11
6) 3. 19	14) 2. 2. 5. 5	22) 7. 11
7) 2. 3. 3. 3	15) 2. 2. 17	23) 2. 3. 3. 5
8) 2. 19	16) 2. 3. 3. 5	24) 2. 37

WWW.MathNotion.Com

ISEE Middle-Level Subject Test Mathematics

Divisibility Rules

1) 8   <u>2</u> 3 <u>4</u> 5 6 7 <u>8</u> 9 10
2) 18   <u>2</u> <u>3</u> 4 5 <u>6</u> 7 8 <u>9</u> 10
3) 55   2 3 4 <u>5</u> 6 7 8 9 10
4) 45   2 <u>3</u> 4 <u>5</u> 6 7 8 <u>9</u> 10
5) 20   <u>2</u> 3 <u>4</u> <u>5</u> 6 7 8 9 <u>10</u>
6) 9   2 <u>3</u> 4 5 6 7 8 <u>9</u> 10
7) 21   2 <u>3</u> 4 5 6 <u>7</u> 8 9 10
8) 28   <u>2</u> 3 <u>4</u> 5 6 <u>7</u> 8 9 10
9) 36   <u>2</u> <u>3</u> <u>4</u> 5 <u>6</u> 7 8 <u>9</u> 10
10) 40   <u>2</u> 3 <u>4</u> <u>5</u> 6 7 <u>8</u> 9 <u>10</u>
11) 39   2 <u>3</u> 4 5 6 7 8 9 10
12) 51   2 <u>3</u> 4 5 6 7 8 9 10

Greatest Common Factor

1) 2
2) 1
3) 3
4) 1
5) 5
6) 8
7) 6
8) 3
9) 6
10) 4
11) 2
12) 4
13) 5
14) 2
15) 9
16) 2
17) 14
18) 8
19) 1
20) 5
21) 1
22) 16
23) 6
24) 9

Least Common Multiple

1) 18
2) 45
3) 80
4) 36
5) 54
6) 42
7) 30
8) 56
9) 21
10) 40
11) 75
12) 63
13) 44
14) 56
15) 56
16) 200
17) 156
18) 22
19) 180
20) 105
21) 162
22) 270
23) 90
24) 75

ISEE Middle-Level Subject Test Mathematics

ISEE Middle-Level Subject Test Mathematics

Chapter 3 :
Fractions

Topics that you will practice in this chapter:

- ✓ Simplifying Fractions
- ✓ Adding and Subtracting Fractions
- ✓ Multiplying and Dividing Fractions
- ✓ Adding and Subtract Mixed Numbers
- ✓ Multiplying and Dividing Mixed Numbers

"A Man is like a fraction whose numerator is what he is and whose denominator is what he thinks of himself. The larger the denominator, the smaller the fraction." −Tolstoy

ISEE Middle-Level Subject Test Mathematics

Simplifying Fractions

✎ **Simplify each fraction to its lowest terms.**

1) $\frac{5}{10} =$

2) $\frac{28}{35} =$

3) $\frac{27}{36} =$

4) $\frac{40}{80} =$

5) $\frac{14}{56} =$

6) $\frac{32}{48} =$

7) $\frac{52}{65} =$

8) $\frac{15}{60} =$

9) $\frac{80}{160} =$

10) $\frac{55}{77} =$

11) $\frac{28}{112} =$

12) $\frac{32}{64} =$

13) $\frac{63}{72} =$

14) $\frac{81}{90} =$

15) $\frac{35}{105} =$

16) $\frac{25}{70} =$

17) $\frac{80}{280} =$

18) $\frac{12}{81} =$

19) $\frac{36}{186} =$

20) $\frac{240}{540} =$

21) $\frac{70}{560} =$

✎ **Find the answer for each problem.**

22) Which of the following fractions equal to $\frac{3}{4}$? ____

    A. $\frac{60}{90}$ B. $\frac{43}{104}$ C. $\frac{48}{64}$ D. $\frac{150}{300}$

23) Which of the following fractions equal to $\frac{5}{8}$? ____

    A. $\frac{125}{200}$ B. $\frac{115}{200}$ C. $\frac{50}{100}$ D. $\frac{30}{90}$

24) Which of the following fractions equal to $\frac{3}{7}$? ____

    A. $\frac{58}{116}$ B. $\frac{54}{126}$ C. $\frac{270}{167}$ D. $\frac{42}{63}$

WWW.MathNotion.Com

ISEE Middle-Level Subject Test Mathematics

Adding and Subtracting Fractions

✎ **Find the sum.**

1) $\frac{5}{9} + \frac{4}{9} =$

2) $\frac{1}{2} + \frac{1}{7} =$

3) $\frac{3}{8} + \frac{1}{4} =$

4) $\frac{3}{5} + \frac{1}{2} =$

5) $\frac{1}{4} + \frac{3}{5} =$

6) $\frac{7}{8} + \frac{3}{8} =$

7) $\frac{1}{2} + \frac{7}{10} =$

8) $\frac{2}{5} + \frac{2}{3} =$

9) $\frac{5}{7} + \frac{2}{3} =$

10) $\frac{7}{12} + \frac{3}{4} =$

11) $\frac{5}{6} + \frac{2}{5} =$

12) $\frac{1}{12} + \frac{2}{3} =$

✎ **Find the difference.**

13) $\frac{1}{3} - \frac{1}{6} =$

14) $\frac{3}{4} - \frac{1}{8} =$

15) $\frac{1}{2} - \frac{1}{3} =$

16) $\frac{1}{4} - \frac{1}{5} =$

17) $\frac{5}{8} - \frac{2}{3} =$

18) $\frac{1}{4} - \frac{1}{7} =$

19) $\frac{5}{6} - \frac{1}{9} =$

20) $\frac{3}{4} - \frac{1}{6} =$

21) $\frac{7}{8} - \frac{1}{12} =$

22) $\frac{8}{15} - \frac{3}{5} =$

23) $\frac{3}{12} - \frac{1}{14} =$

24) $\frac{10}{13} - \frac{7}{26} =$

25) $\frac{6}{7} - \frac{3}{4} =$

26) $\frac{4}{5} - \frac{1}{8} =$

27) $\frac{4}{7} - \frac{2}{35} =$

28) $\frac{9}{16} - \frac{2}{8} =$

29) $\frac{8}{9} - \frac{7}{18} =$

30) $\frac{1}{2} - \frac{4}{9} =$

WWW.MathNotion.Com

ISEE Middle-Level Subject Test Mathematics

Multiplying and Dividing Fractions

✍ Find the value of each expression in lowest terms.

1) $\frac{1}{5} \times \frac{15}{5} =$

2) $\frac{9}{12} \times \frac{4}{9} =$

3) $\frac{1}{16} \times \frac{8}{10} =$

4) $\frac{1}{24} \times \frac{8}{10} =$

5) $\frac{1}{5} \times \frac{1}{4} =$

6) $\frac{7}{9} \times \frac{1}{7} =$

7) $\frac{6}{7} \times \frac{1}{3} =$

8) $\frac{2}{8} \times \frac{2}{8} =$

9) $\frac{5}{8} \times \frac{3}{5} =$

10) $\frac{4}{7} \times \frac{1}{8} =$

11) $\frac{7}{15} \times \frac{5}{7} =$

12) $\frac{3}{10} \times \frac{5}{9} =$

✍ Find the value of each expression in lowest terms.

13) $\frac{1}{4} \div \frac{1}{8} =$

14) $\frac{1}{10} \div \frac{1}{5} =$

15) $\frac{3}{4} \div \frac{1}{5} =$

16) $\frac{1}{3} \div \frac{5}{6} =$

17) $\frac{1}{7} \div \frac{8}{42} =$

18) $\frac{3}{4} \div \frac{1}{6} =$

19) $\frac{2}{7} \div \frac{7}{13} =$

20) $\frac{1}{24} \div \frac{3}{16} =$

21) $\frac{7}{12} \div \frac{5}{6} =$

22) $\frac{22}{18} \div \frac{11}{9} =$

23) $\frac{9}{35} \div \frac{3}{7} =$

24) $\frac{2}{7} \div \frac{8}{21} =$

25) $\frac{1}{9} \div \frac{2}{5} =$

26) $\frac{5}{12} \div \frac{3}{5} =$

27) $\frac{3}{20} \div \frac{1}{6} =$

28) $\frac{8}{20} \div \frac{3}{4} =$

29) $\frac{5}{6} \div \frac{2}{9} =$

30) $\frac{5}{11} \div \frac{3}{4} =$

WWW.MathNotion.Com

ISEE Middle-Level Subject Test Mathematics

Adding and Subtracting Mixed Numbers

✎ **Find the sum.**

1) $3\frac{1}{3} + 2\frac{1}{6} =$

2) $4\frac{1}{2} + 3\frac{1}{2} =$

3) $3\frac{3}{8} + 1\frac{1}{8} =$

4) $2\frac{1}{4} + 2\frac{1}{3} =$

5) $3\frac{5}{6} + 2\frac{7}{12} =$

6) $5\frac{4}{15} + 3\frac{3}{5} =$

7) $2\frac{1}{3} + 4\frac{3}{7} =$

8) $3\frac{1}{2} + 4\frac{2}{5} =$

9) $5\frac{2}{5} + 6\frac{3}{7} =$

10) $8\frac{5}{16} + 6\frac{1}{12} =$

✎ **Find the difference.**

11) $3\frac{1}{4} - 1\frac{3}{4} =$

12) $6\frac{3}{5} - 4\frac{2}{5} =$

13) $4\frac{1}{3} - 3\frac{1}{9} =$

14) $7\frac{1}{7} - 5\frac{1}{2} =$

15) $5\frac{1}{3} - 2\frac{1}{12} =$

16) $8\frac{1}{5} - 4\frac{1}{3} =$

17) $9\frac{1}{4} - 6\frac{1}{8} =$

18) $11\frac{7}{15} - 8\frac{3}{5} =$

19) $14\frac{5}{6} - 11\frac{3}{5} =$

20) $18\frac{2}{7} - 14\frac{1}{5} =$

21) $9\frac{1}{3} - 4\frac{1}{4} =$

22) $6\frac{1}{8} - 4\frac{1}{16} =$

23) $19\frac{3}{8} - 15\frac{1}{3} =$

24) $11\frac{1}{9} - 8\frac{1}{8} =$

25) $17\frac{1}{7} - 11\frac{1}{5} =$

26) $16\frac{2}{9} - 9\frac{5}{7} =$

WWW.MathNotion.Com

ISEE Middle-Level Subject Test Mathematics

Multiplying and Dividing Mixed Numbers

✏ **Find the product.**

1) $5\frac{1}{2} \times 2\frac{1}{4} =$

2) $5\frac{1}{3} \times 4\frac{1}{3} =$

3) $5\frac{3}{4} \times 6\frac{1}{4} =$

4) $3\frac{1}{3} \times 2\frac{3}{5} =$

5) $4\frac{8}{10} \times 1\frac{1}{24} =$

6) $6\frac{2}{7} \times 1\frac{1}{11} =$

7) $8\frac{2}{3} \times 3\frac{1}{2} =$

8) $3\frac{4}{7} \times 2\frac{1}{5} =$

9) $5\frac{2}{8} \times 4\frac{1}{6} =$

10) $7\frac{3}{3} \times 1\frac{3}{8} =$

✏ **Find the quotient.**

11) $2\frac{2}{5} \div 4\frac{1}{5} =$

12) $4\frac{1}{6} \div 3\frac{1}{3} =$

13) $6\frac{1}{3} \div 1\frac{1}{2} =$

14) $7\frac{1}{10} \div 2\frac{2}{5} =$

15) $3\frac{1}{3} \div 1\frac{1}{9} =$

16) $1\frac{1}{10} \div 4\frac{1}{2} =$

17) $1\frac{3}{16} \div 5\frac{1}{4} =$

18) $4\frac{1}{3} \div 4\frac{3}{4} =$

19) $9\frac{1}{3} \div 2\frac{1}{4} =$

20) $15\frac{1}{3} \div 5\frac{1}{2} =$

21) $4\frac{1}{6} \div 1\frac{1}{5} =$

22) $1\frac{1}{18} \div 1\frac{2}{9} =$

23) $4\frac{2}{7} \div 1\frac{3}{10} =$

24) $7\frac{1}{3} \div 2\frac{2}{11} =$

25) $8\frac{2}{5} \div 1\frac{1}{6} =$

26) $9\frac{1}{3} \div 2\frac{1}{7} =$

WWW.MathNotion.Com

ISEE Middle-Level Subject Test Mathematics

Answers of Worksheets

Simplifying Fractions

1) $\frac{1}{2}$
2) $\frac{4}{5}$
3) $\frac{3}{4}$
4) $\frac{1}{2}$
5) $\frac{1}{4}$
6) $\frac{2}{3}$
7) $\frac{4}{5}$
8) $\frac{1}{4}$
9) $\frac{1}{2}$
10) $\frac{5}{7}$
11) $\frac{1}{4}$
12) $\frac{1}{2}$
13) $\frac{7}{8}$
14) $\frac{9}{10}$
15) $\frac{1}{3}$
16) $\frac{5}{14}$
17) $\frac{2}{7}$
18) $\frac{4}{27}$
19) $\frac{6}{31}$
20) $\frac{4}{9}$
21) $\frac{1}{8}$
22) C
23) A
24) B

Adding and Subtracting Fractions

1) $\frac{9}{9} = 1$
2) $\frac{9}{14}$
3) $\frac{5}{8}$
4) $1\frac{1}{10}$
5) $\frac{17}{20}$
6) $1\frac{1}{4}$
7) $1\frac{1}{5}$
8) $1\frac{1}{15}$
9) $1\frac{8}{21}$
10) $1\frac{1}{3}$
11) $1\frac{7}{30}$
12) $\frac{3}{4}$
13) $\frac{1}{6}$
14) $\frac{5}{8}$
15) $\frac{1}{6}$
16) $\frac{1}{20}$
17) $-\frac{1}{24}$
18) $\frac{3}{28}$
19) $\frac{13}{18}$
20) $\frac{7}{12}$
21) $\frac{19}{24}$
22) $-\frac{1}{15}$
23) $\frac{5}{28}$
24) $\frac{1}{2}$
25) $\frac{3}{28}$
26) $\frac{27}{40}$
27) $\frac{18}{35}$
28) $\frac{5}{16}$
29) $\frac{1}{2}$
30) $\frac{1}{18}$

Multiplying and Dividing Fractions

1) $\frac{3}{5}$
2) $\frac{1}{3}$
3) $\frac{1}{20}$
4) $\frac{1}{30}$
5) $\frac{1}{20}$
6) $\frac{1}{9}$
7) $\frac{2}{7}$
8) $\frac{1}{16}$
9) $\frac{3}{8}$
10) $\frac{1}{14}$
11) $\frac{1}{3}$
12) $\frac{1}{6}$
13) 2
14) $\frac{1}{2}$
15) $3\frac{3}{4}$
16) $\frac{2}{5}$

WWW.MathNotion.Com

ISEE Middle-Level Subject Test Mathematics

17) $\frac{3}{4}$

18) $4\frac{1}{2}$

19) $\frac{26}{49}$

20) $\frac{2}{9}$

21) $\frac{7}{10}$

22) 1

23) $\frac{3}{5}$

24) $\frac{3}{4}$

25) $\frac{5}{18}$

26) $\frac{25}{36}$

27) $\frac{9}{10}$

28) $\frac{8}{15}$

29) $3\frac{3}{4}$

30) $\frac{20}{33}$

Adding and Subtracting Mixed Numbers

1) $5\frac{1}{2}$

2) 8

3) $4\frac{1}{2}$

4) $4\frac{7}{12}$

5) $6\frac{5}{12}$

6) $8\frac{13}{15}$

7) $6\frac{16}{21}$

8) $7\frac{9}{10}$

9) $11\frac{29}{35}$

10) $14\frac{19}{48}$

11) $1\frac{1}{2}$

12) $2\frac{1}{5}$

13) $1\frac{2}{9}$

14) $1\frac{9}{14}$

15) $3\frac{1}{4}$

16) $3\frac{13}{15}$

17) $3\frac{1}{8}$

18) $2\frac{13}{15}$

19) $3\frac{7}{30}$

20) $4\frac{3}{35}$

21) $5\frac{1}{12}$

22) $2\frac{1}{16}$

23) $4\frac{1}{24}$

24) $2\frac{71}{72}$

25) $5\frac{33}{35}$

26) $6\frac{32}{63}$

Multiplying and Dividing Mixed Numbers

1) $12\frac{3}{8}$

2) $23\frac{1}{9}$

3) $35\frac{15}{16}$

4) $8\frac{2}{3}$

5) 5

6) $6\frac{6}{7}$

7) $30\frac{1}{3}$

8) $7\frac{6}{7}$

9) $21\frac{7}{8}$

10) 11

11) $\frac{4}{7}$

12) $1\frac{1}{4}$

13) $4\frac{2}{9}$

14) $2\frac{23}{24}$

15) 3

16) $\frac{11}{45}$

17) $\frac{19}{84}$

18) $\frac{52}{57}$

19) $4\frac{4}{27}$

20) $2\frac{26}{33}$

21) $3\frac{17}{36}$

22) $\frac{19}{22}$

23) $3\frac{27}{91}$

24) $3\frac{13}{36}$

25) $7\frac{1}{5}$

26) $4\frac{16}{45}$

ISEE Middle-Level Subject Test Mathematics

Chapter 4:
Decimals

Topics that you will practice in this chapter:

- ✓ Adding and Subtracting Decimals
- ✓ Multiplying and Dividing Decimals
- ✓ Comparing Decimals
- ✓ Rounding Decimals

*"The study of mathematics, like the Nile, begins in minuteness but ends in magnificence." –
Charles Caleb Colton*

ISEE Middle-Level Subject Test Mathematics

Adding and Subtracting Decimals

✎ Add and subtract decimals.

1) 35.19 − 24.28 = _____

2) 34.29 + 42.58 = _____

3) 61.20 + 33.75 = _____

4) 38.72 − 21.68 = _____

5) 57.39 + 26.54 = _____

6) 70.24 − 42.35 = _____

7) 86.09 − 35.14 = _____

8) 54.51 + 32.66 = _____

9) 114.21 − 88.69 = _____

✎ Find the missing number.

10) ___ + 2.8 = 5.4

11) 4.1 + ___ = 5.88

12) 6.45 + ___ = 8

13) 7.25 − ___ = 3.40

14) ___ − 2.35 = 4.25

15) ___ − 19.85 = 6.54

16) 22.15 + ___ = 28.95

17) ___ − 37.16 = 9.42

18) ___ + 24.50 = 34.19

19) 72.40 + ___ = 125.20

ISEE Middle-Level Subject Test Mathematics

Multiplying and Dividing Decimals

✏ **Find the product.**

1) 0.5 × 0.6 =

2) 3.3 × 0.4 =

3) 1.28 × 0.5 =

4) 0.35 × 0.6 =

5) 1.85 × 0.6 =

6) 0.24 × 0.5 =

7) 5.25 × 1.4 =

8) 18.5 × 4.6 =

9) 15.4 × 6.8 =

10) 19.5 × 2.6 =

11) 32.2 × 1.5 =

12) 78.4 × 4.5 =

✏ **Find the quotient.**

13) 1.85 ÷ 10 =

14) 74.6 ÷ 100 =

15) 3.6 ÷ 3 =

16) 9.6 ÷ 0.4 =

17) 15.5 ÷ 0.5 =

18) 32.8 ÷ 0.2 =

19) 22.15 ÷ 1,000 =

20) 53.55 ÷ 0.7 =

21) 322.2 ÷ 0.2 =

22) 50.67 ÷ 0.18 =

23) 77.4 ÷ 0.8 =

24) 27.93 ÷ 0.03 =

ISEE Middle-Level Subject Test Mathematics

Comparing Decimals

✎ **Write the correct comparison symbol (>, < or =).**

1) 0.70 ☐ 0.070

2) 0.049 ☐ 0.49

3) 5.090 ☐ 5.09

4) 2.57 ☐ 2.05

5) 9.03 ☐ 0.930

6) 6.06 ☐ 6.6

7) 7.02 ☐ 7.020

8) 3.04 ☐ 3.2

9) 3.61 ☐ 3.245

10) 0.986 ☐ 0.0986

11) 17.24 ☐ 17.240

12) 0.759 ☐ 0.81

13) 9.040 ☐ 9.40

14) 5.73 ☐ 5.213

15) 9.44 ☐ 9.404

16) 7.17 ☐ 7.170

17) 4.85 ☐ 4.085

18) 9.041 ☐ 9.40

19) 3.033 ☐ 3.030

20) 4.97 ☐ 4.970

WWW.MathNotion.Com

ISEE Middle-Level Subject Test Mathematics

Rounding Decimals

✍ **Round each decimal to the nearest whole number.**

1) 28.12   3) 16.22   5) 7.95

2) 6.9   4) 8.5   6) 52.7

✍ **Round each decimal to the nearest tenth.**

7) 31.761   9) 94.729   11) 13.219

8) 14.421   10) 77.89   12) 59.89

✍ **Round each decimal to the nearest hundredth.**

13) 8.428   15) 55.3786   17) 62.241

14) 23.812   16) 231.912   18) 19.447

✍ **Round each decimal to the nearest thousandth.**

19) 15.54324   21) 243.8652   23) 67.1983

20) 34.62586   22) 80.4529   24) 72.36788

ISEE Middle-Level Subject Test Mathematics

Convert Fraction to Decimal

✎ Write each as a decimal.

1) $\dfrac{50}{100} =$

2) $\dfrac{46}{100} =$

3) $\dfrac{8}{50} =$

4) $\dfrac{8}{32} =$

5) $\dfrac{8}{72} =$

6) $\dfrac{56}{100} =$

7) $\dfrac{4}{50} =$

8) $\dfrac{31}{48} =$

9) $\dfrac{27}{300} =$

10) $\dfrac{15}{55} =$

11) $\dfrac{16}{32} =$

12) $\dfrac{6}{16} =$

13) $\dfrac{3}{10} =$

14) $\dfrac{18}{250} =$

15) $\dfrac{24}{80} =$

16) $\dfrac{30}{40} =$

17) $\dfrac{68}{100} =$

18) $\dfrac{7}{35} =$

19) $\dfrac{87}{100} =$

20) $\dfrac{1}{100} =$

21) $\dfrac{6}{36} =$

22) $\dfrac{2}{80} =$

WWW.MathNotion.Com

ISEE Middle-Level Subject Test Mathematics

Convert Decimal to Percent

✎ Write each as a percent.

1) 0.187 =

2) 0.19 =

3) 2.6 =

4) 0.017 =

5) 0.009 =

6) 0.786 =

7) 0.245 =

8) 0.57 =

9) 0.002 =

10) 0.205 =

11) 0.324 =

12) 84.9 =

13) 3.015 =

14) 0.7 =

15) 2.35 =

16) 0.0367 =

17) 0.0043 =

18) 0.960 =

19) 6.68 =

20) 0.484 =

21) 8.957 =

22) 0.879 =

23) 2.7 =

24) 0.9 =

25) 3.6 =

26) 26.8 =

27) 1.01 =

28) 0.006 =

ISEE Middle-Level Subject Test Mathematics

Convert Fraction to Percent

✎ **Write each as a percent.**

1) $\dfrac{1}{4} =$

2) $\dfrac{3}{8} =$

3) $\dfrac{7}{14} =$

4) $\dfrac{15}{35} =$

5) $\dfrac{12}{28} =$

6) $\dfrac{17}{68} =$

7) $\dfrac{8}{11} =$

8) $\dfrac{14}{30} =$

9) $\dfrac{6}{50} =$

10) $\dfrac{12}{48} =$

11) $\dfrac{5}{34} =$

12) $\dfrac{27}{10} =$

13) $\dfrac{24}{80} =$

14) $\dfrac{16}{25} =$

15) $\dfrac{16}{58} =$

16) $\dfrac{2}{22} =$

17) $\dfrac{32}{88} =$

18) $\dfrac{21}{36} =$

19) $\dfrac{18}{92} =$

20) $\dfrac{6}{60} =$

21) $\dfrac{24}{600} =$

22) $\dfrac{720}{360} =$

ISEE Middle-Level Subject Test Mathematics

Answers of Worksheets

Adding and Subtracting Decimals

1) 10.91
2) 76.87
3) 94.95
4) 17.04
5) 83.93
6) 27.89
7) 50.95
8) 87.17
9) 25.52
10) 2.6
11) 1.78
12) 1.55
13) 3.85
14) 6.6
15) 26.39
16) 6.8
17) 46.58
18) 9.69
19) 52.8

Multiplying and Dividing Decimals

1) 0.3
2) 1.32
3) 0.64
4) 0.21
5) 1.11
6) 0.12
7) 7.35
8) 85.1
9) 104.72
10) 50.7
11) 48.3
12) 352.8
13) 0.185
14) 0.746
15) 1.2
16) 24
17) 31
18) 164
19) 0.02215
20) 76.5
21) 1,611
22) 281.5
23) 96.75
24) 931

Comparing Decimals

1) >
2) <
3) =
4) >
5) >
6) <
7) =
8) <
9) >
10) >
11) =
12) <
13) <
14) >
15) >
16) =
17) >
18) <
19) >
20) =

Rounding Decimals

1) 28
2) 7
3) 16
4) 9
5) 8
6) 53
7) 31.8
8) 14.4
9) 94.7
10) 77.9
11) 13.2
12) 59.9
13) 8.43
14) 23.81
15) 55.38
16) 231.91
17) 62.24
18) 19.45
19) 15.543
20) 34.626
21) 243.865
22) 80.453
23) 67.198
24) 72.368

Convert Fraction to Decimal

1) 0.5
2) 0.46
3) 0.16

WWW.MathNotion.Com

ISEE Middle-Level Subject Test Mathematics

4) 0.25
5) 0.11
6) 0.56
7) 0.08
8) 0.646
9) 0.09
10) 0.27

11) 0.5
12) 0.375
13) 0.3
14) 0.072
15) 0.3
16) 0.75
17) 0.68

18) 0.2
19) 0.87
20) 0.01
21) 0.166
22) 0.025

Convert Decimal to Percent

1) 18.7%
2) 19%
3) 260%
4) 1.7%
5) 0.9%
6) 78.6%
7) 24.5%
8) 57%
9) 0.2%
10) 20.5%

11) 32.4%
12) 8,490%
13) 301.5%
14) 70%
15) 235%
16) 3.67%
17) 0.43%
18) 96%
19) 668%
20) 48.4%

21) 895.7%
22) 87.9%
23) 270%
24) 90%
25) 360%
26) 2,680%
27) 101%
28) 0.6%

Convert Fraction to Percent

1) 25%
2) 37.5%
3) 50%
4) 42.86%
5) 29.31%
6) 25%
7) 72.72%
8) 46.66%

9) 12%
10) 25%
11) 14.7%
12) 2.7%
13) 30%
14) 64%
15) 27.58%
16) 9.09%

17) 36.36%
18) 58.33%
19) 19.56%
20) 10%
21) 4%
22) 200%

ISEE Middle-Level Subject Test Mathematics

Chapter 5:
Proportions, Ratios, and Percent

Topics that you will practice in this chapter:

- ✓ Simplifying Ratios
- ✓ Proportional Ratios
- ✓ Similarity and Ratios
- ✓ Ratio and Rates Word Problems
- ✓ Percentage Calculations
- ✓ Percent Problems
- ✓ Discount, Tax and Tip

Without mathematics, there's nothing you can do. Everything around you is mathematics. Everything around you is numbers." – Shakuntala Devi

ISEE Middle-Level Subject Test Mathematics

Simplifying Ratios

✎ Reduce each ratio.

1) $15:20 = \underline{}:\underline{}$

2) $7:70 = \underline{}:\underline{}$

3) $16:28 = \underline{}:\underline{}$

4) $7:21 = \underline{}:\underline{}$

5) $4:40 = \underline{}:\underline{}$

6) $6:48 = \underline{}:\underline{}$

7) $16:64 = \underline{}:\underline{}$

8) $10:25 = \underline{}:\underline{}$

9) $8:48 = \underline{}:\underline{}$

10) $49:63 = \underline{}:\underline{}$

11) $18:27 = \underline{}:\underline{}$

12) $35:10 = \underline{}:\underline{}$

13) $90:9 = \underline{}:\underline{}$

14) $24:32 = \underline{}:\underline{}$

15) $7:56 = \underline{}:\underline{}$

16) $45:63 = \underline{}:\underline{}$

17) $56:72 = \underline{}:\underline{}$

18) $26:13 = \underline{}:\underline{}$

19) $15:45 = \underline{}:\underline{}$

20) $28:4 = \underline{}:\underline{}$

21) $24:48 = \underline{}:\underline{}$

22) $30:24 = \underline{}:\underline{}$

23) $70:140 = \underline{}:\underline{}$

24) $6:180 = \underline{}:\underline{}$

✎ Write each ratio as a fraction in simplest form.

25) $6:12 =$

26) $30:50 =$

27) $15:35 =$

28) $9:27 =$

29) $8:24 =$

30) $18:84 =$

31) $7:14 =$

32) $7:35 =$

33) $40:96 =$

34) $12:54 =$

35) $44:52 =$

36) $12:27 =$

37) $15:180 =$

38) $39:143 =$

39) $20:300 =$

40) $30:120 =$

41) $56:42 =$

42) $26:130 =$

43) $66:123 =$

44) $70:630 =$

45) $75:125 =$

WWW.MathNotion.Com

ISEE Middle-Level Subject Test Mathematics

Proportional Ratios

✎ Fill in the blanks; Calculate each proportion.

1) $3:8 = \_\_ : 48$

2) $2:5 = 20:\_\_$

3) $1:9 = \_\_ : 81$

4) $6:7 = 12:\_\_$

5) $9:2 = 63:\_\_$

6) $8:7 = \_\_ : 49$

7) $20:3 = \_\_ : 15$

8) $1:3 = \_\_ : 75$

9) $7:6 = \_\_ : 60$

10) $8:5 = \_\_ : 45$

11) $3:10 = 60:\_\_$

12) $6:11 = 42:\_\_$

✎ State if each pair of ratios form a proportion.

13) $\frac{3}{20}$ and $\frac{9}{60}$

14) $\frac{1}{7}$ and $\frac{6}{42}$

15) $\frac{3}{7}$ and $\frac{24}{56}$

16) $\frac{4}{9}$ and $\frac{12}{18}$

17) $\frac{1}{9}$ and $\frac{12}{81}$

18) $\frac{7}{8}$ and $\frac{21}{28}$

19) $\frac{9}{13}$ and $\frac{27}{39}$

20) $\frac{1}{8}$ and $\frac{8}{64}$

21) $\frac{6}{19}$ and $\frac{30}{85}$

22) $\frac{5}{9}$ and $\frac{40}{81}$

23) $\frac{9}{14}$ and $\frac{108}{168}$

24) $\frac{15}{23}$ and $\frac{360}{552}$

✎ Calculate each proportion.

25) $\frac{20}{25} = \frac{32}{x}, x = \_\_\_$

26) $\frac{1}{8} = \frac{32}{x}, x = \_\_\_$

27) $\frac{15}{5} = \frac{21}{x}, x = \_\_\_$

28) $\frac{1}{7} = \frac{x}{294}, x = \_\_\_$

29) $\frac{7}{9} = \frac{x}{81}, x = \_\_\_$

30) $\frac{1}{5} = \frac{13}{x}, x = \_\_\_$

31) $\frac{9}{5} = \frac{36}{x}, x = \_\_\_$

32) $\frac{6}{13} = \frac{48}{x}, x = \_\_\_$

33) $\frac{5}{8} = \frac{x}{88}, x = \_\_\_$

34) $\frac{4}{15} = \frac{x}{240}, x = \_\_\_$

35) $\frac{9}{19} = \frac{x}{266}, x = \_\_\_$

36) $\frac{7}{15} = \frac{x}{270}, x = \_\_\_$

WWW.MathNotion.Com

ISEE Middle-Level Subject Test Mathematics

Similarity and Ratios

✏️ **Each pair of figures is similar. Find the missing side.**

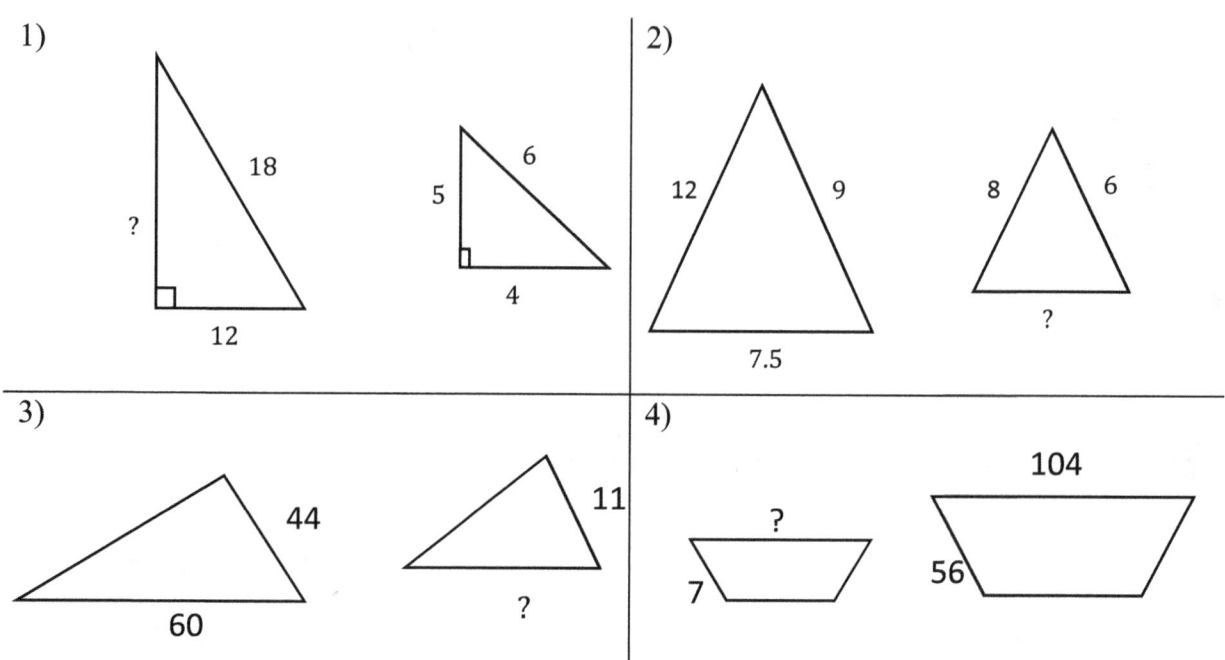

✏️ **Calculate.**

5) Two rectangles are similar. The first is 24 feet wide and 120 feet long. The second is 30 feet wide. What is the length of the second rectangle? _____

6) Two rectangles are similar. One is 5 meters by 36 meters. The longer side of the second rectangle is 90 meters. What is the other side of the second rectangle? _____

7) A building casts a shadow 25 ft long. At the same time a girl 10 ft tall casts a shadow 5 ft long. How tall is the building? _____

8) The scale of a map of Texas is 4 inches: 32 miles. If you measure the distance from Dallas to Martin County as 38.4 inches, approximately how far is Martin County from Dallas? _____

ISEE Middle-Level Subject Test Mathematics

Ratio and Rates Word Problems

✍ **Find the answer for each word problem.**

1) Mason has 24 red cards and 36 green cards. What is the ratio of Mason's red cards to his green cards? _____

2) In a party, 45 soft drinks are required for every 54 guests. If there are 378 guests, how many soft drinks is required? _____

3) In Mason's class, 42 of the students are tall and 24 are short. In Michael's class 84 students are tall and 48 students are short. Which class has a higher ratio of tall to short students? _____

4) The price of 5 apples at the Quick Market is $4.6. The price of 7 of the same apples at Walmart is $5.95. Which place is the better buy? _____

5) The bakers at a Bakery can make 90 bagels in 3 hours. How many bagels can they bake in 24 hours? What is that rate per hour? _____

6) You can buy 5 cans of green beans at a supermarket for $5.75. How much does it cost to buy 45 cans of green beans? _____

7) The ratio of boys to girls in a class is 4: 7. If there are 32 boys in the class, how many girls are in that class? _____

8) The ratio of red marbles to blue marbles in a bag is 3: 7. If there are 50 marbles in the bag, how many of the marbles are red? _____

ISEE Middle-Level Subject Test Mathematics

Percentage Calculations

✏ **Calculate the given percent of each value.**

1) 3% of 60 = ___

2) 20% of 32 = ___

3) 4% of 72 = ___

4) 16% of 32 = ___

5) 25% of 124 = ___

6) 35% of 56 = ___

7) 15% of 20 = ___

8) 14% of 150 = ___

9) 80% of 50 = ___

10) 12% of 115 = ___

11) 72% of 250 = ___

12) 52% of 500 = ___

13) 70% of 400 = ___

14) 27% of 145 = ___

15) 90% of 64 = ___

16) 60% of 55 = ___

17) 22% of 210 = ___

18) 8% of 235 = ___

✏ **Calculate the percent of each given value.**

19) ___% of 25 = 5

20) ___% of 40 = 20

21) ___% of 25 = 2

22) ___% of 50 = 16

23) ___% of 250 = 5

24) ___% of 40 = 32

25) ___% of 125 = 20

26) ___% of 700 = 49

27) ___% of 350 = 49

28) ___% of 500 = 210

✏ **Calculate each percent problem.**

29) A Cinema has 250 seats. 60 seats were sold for the current movie. What percent of seats are empty? ____ %

30) There are 68 boys and 92 girls in a class. 75% of the students in the class take the bus to school. How many students do not take the bus to school? ____

WWW.MathNotion.Com

ISEE Middle-Level Subject Test Mathematics

Percent Problems

✏ Calculate each problem.

1) 9 is what percent of 45? ___%

2) 60 is what percent of 120? ___%

3) 10 is what percent of 200? ___%

4) 15 is what percent of 125? ___%

5) 10 is what percent of 400? ___%

6) 66 is what percent of 55? ___%

7) 40 is what percent of 160? ___%

8) 40 is what percent of 50? ___%

9) 120 is what percent of 800? ___%

10) 78 is what percent of 120? ___%

11) 36 is what percent of 144? ___%

12) 17 is what percent of 85? ___%

13) 90 is what percent of 900? ___%

14) 36 is what percent of 16? ___%

15) 63 is what percent of 14? ___%

16) 18 is what percent of 60? ___%

17) 126 is what percent of 200? ___%

18) 232 is what percent of 40? ___%

✏ Calculate each percent word problem.

19) There are 40 employees in a company. On a certain day, 25 were present. What percent showed up for work? ___%

20) A metal bar weighs 60 ounces. 25% of the bar is gold. How many ounces of gold are in the bar? _____

21) A crew is made up of 12 women; the rest are men. If 15% of the crew are women, how many people are in the crew? _____

22) There are 40 students in a class and 8 of them are girls. What percent are boys? ___%

23) The Royals softball team played 400 games and won 280 of them. What percent of the games did they lose? ___%

WWW.MathNotion.Com

ISEE Middle-Level Subject Test Mathematics

Discount, Tax and Tip

🪶 **Find the selling price of each item.**

1) Original price of a computer: $420
Tax: 8% Selling price: $_____

2) Original price of a laptop: $280
Tax: 4% Selling price: $_____

3) Original price of a sofa: $820
Tax: 5% Selling price: $_____

4) Original price of a car: $15,800
Tax: 3.6% Selling price: $_____

5) Original price of a Table: $250
Tax: 9% Selling price: $_____

6) Original price of a house: $630,000
Tax: 1.8% Selling price: $_____

7) Original price of a tablet: $450
Discount: 30% Selling price: $____

8) Original price of a chair: $390
Discount: 8% Selling price: $____

9) Original price of a book: $75
Discount: 42% Selling price: $____

10) Original price of a cellphone: $820
Discount: 23% Selling price: $___

11) Food bill: $45
Tip: 15% Price: $_____

12) Food bill: $32
Tipp: 20% Price: $_____

13) Food bill: $90
Tip: 35% Price: $_____

14) Food bill: $42
Tipp: 12% Price: $_____

🪶 **Find the answer for each word problem.**

15) Nicolas hired a moving company. The company charged $500 for its services, and Nicolas gives the movers a 40% tip. How much does Nicolas tip the movers? $_____

16) Mason has lunch at a restaurant and the cost of his meal is $90. Mason wants to leave a 25% tip. What is Mason's total bill including tip? $_____

17) The sales tax in Texas is 19.80% and an item costs $350. How much is the tax? $_____

18) The price of a table at Best Buy is $680. If the sales tax is 5%, what is the final price of the table including tax? $_____

ISEE Middle-Level Subject Test Mathematics

Answers of Worksheets

Simplifying Ratios

1) 3 : 4
2) 1 : 10
3) 4 : 7
4) 1 : 3
5) 1 : 10
6) 1 : 8
7) 2 : 8
8) 2 : 5
9) 1 : 6
10) 7 : 9
11) 2 : 3
12) 7 : 2
13) 10 : 1
14) 3 : 4
15) 1 : 8
16) 5 : 7
17) 7 : 9
18) 2 : 1
19) 1 : 3
20) 7 : 1
21) 1 : 2
22) 5 : 4
23) 1 : 2
24) 1 : 30
25) $\frac{1}{2}$
26) $\frac{3}{5}$
27) $\frac{3}{7}$
28) $\frac{1}{3}$
29) $\frac{1}{3}$
30) $\frac{3}{14}$
31) $\frac{1}{2}$
32) $\frac{1}{5}$
33) $\frac{5}{12}$
34) $\frac{2}{9}$
35) $\frac{11}{13}$
36) $\frac{4}{9}$
37) $\frac{1}{12}$
38) $\frac{3}{11}$
39) $\frac{1}{15}$
40) $\frac{1}{4}$
41) $\frac{4}{3}$
42) $\frac{1}{5}$
43) $\frac{22}{41}$
44) $\frac{1}{9}$
45) $\frac{3}{5}$

Proportional Ratios

1) 18
2) 50
3) 9
4) 14
5) 14
6) 56
7) 100
8) 25
9) 70
10) 72
11) 200
12) 77
13) Yes
14) Yes
15) Yes
16) No
17) No
18) No
19) Yes
20) Yes
21) No
22) No
23) Yes
24) Yes
25) 40
26) 256
27) 7
28) 42
29) 63
30) 65
31) 20
32) 104
33) 55
34) 64
35) 126
36) 126

Similarity and ratios

1) 15
2) 5
3) 15
4) 13
5) 150 feet
6) 12.5 meters
7) 50 feet
8) 307.2 miles

Ratio and Rates Word Problems

1) 2 : 3
2) 315

WWW.MathNotion.Com

ISEE Middle-Level Subject Test Mathematics

3) The ratio for both classes is 7 to 4.
4) Walmart is a better buy.
5) 720, the rate is 30 per hour.

6) $51.75
7) 56
8) 15

Percentage Calculations

1) 1.8	11) 180	21) 8%
2) 6.4	12) 260	22) 32%
3) 2.88	13) 280	23) 2%
4) 5.12	14) 39.15	24) 80%
5) 31	15) 57.6	25) 16%
6) 19.6	16) 33	26) 7%
7) 3	17) 46.2	27) 14%
8) 21	18) 18.8	28) 42%
9) 40	19) 20%	29) 76%
10) 13.8	20) 50%	30) 40

Percent Problems

1) 20%	9) 15%	17) 63%
2) 50%	10) 65%	18) 580%
3) 5%	11) 25%	19) 62.5%
4) 12%	12) 20%	20) 15 ounces
5) 2.5%	13) 10%	21) 80
6) 120%	14) 225%	22) 80%
7) 25%	15) 450%	23) 30%
8) 80%	16) 30%	

Discount, Tax and Tip

1) $453.60	7) $315.00	13) $121.50
2) $291.20	8) $358.80	14) $47.04
3) $861.00	9) $43.50	15) $200.00
4) $16,368.80	10) $631.40	16) $112.50
5) $272.50	11) $51.75	17) $69.30
6) $641,340	12) $38.40	18) $714.00

WWW.MathNotion.Com

ISEE Middle-Level Subject Test Mathematics

Chapter 6 :
Exponents and Radicals Expressions

Topics that you will practice in this chapter:

- ✓ Adding and Subtracting Exponents
- ✓ Multiplication Property of Exponents
- ✓ Zero and Negative Exponents
- ✓ Division Property of Exponents
- ✓ Powers of Products and Quotients
- ✓ Negative Exponents and Negative Bases
- ✓ Scientific Notation
- ✓ Square Roots

Mathematics is no more computation than typing is literature.

– John Allen Paulos

ISEE Middle-Level Subject Test Mathematics

Adding and Subtracting Exponents

✏️ Solve each problem.

1) $3^2 + 2^5 =$

2) $x^6 + x^6 =$

3) $3b^2 - 2b^2 =$

4) $3 + 4^3 =$

5) $8 - 4^2 =$

6) $4 + 7^1 =$

7) $2x^3 + 3x^3 =$

8) $10^2 + 3^5 =$

9) $4^5 - 2^4 =$

10) $5^2 - 6^0 =$

11) $1^2 - 3^0 =$

12) $7^1 + 2^3 =$

13) $6^1 - 5^3 =$

14) $3^3 + 3^3 =$

15) $9^2 - 8^2 =$

16) $0^{73} + 0^{54} =$

17) $2^2 - 3^2 =$

18) $7^3 - 7^1 =$

19) $8^2 - 6^2 =$

20) $4^2 + 3^2 =$

21) $2^3 + 4^3 =$

22) $10 + 3^3 =$

23) $6x^5 + 8x^5 =$

24) $8^0 + 4^2 =$

25) $3^2 + 3^2 =$

26) $10^2 + 5^2 =$

27) $(\frac{1}{2})^2 + (\frac{1}{2})^2 =$

28) $9^2 + 3^2 =$

WWW.MathNotion.Com

ISEE Middle-Level Subject Test Mathematics

Multiplication Property of Exponents

✎ **Simplify and write the answer in exponential form.**

1) $4 \times 4^5 =$

2) $8^4 \times 8 =$

3) $7^3 \times 7^3 =$

4) $9^2 \times 9^2 =$

5) $2^2 \times 2^4 \times 2 =$

6) $5 \times 5^3 \times 5^3 =$

7) $4^3 \times 4^2 \times 4 \times 4 =$

8) $5x \times x =$

9) $x^3 \times x^3 =$

10) $x^7 \times x^2 =$

11) $x^4 \times x^3 \times x^2 =$

12) $10x \times 3x =$

13) $4x^3 \times 4x^3 =$

14) $7x^3 \times x =$

15) $3x^2 \times 4x^2 \times x^2 =$

16) $5x^4 \times x^4 =$

17) $2x^8 \times 2x =$

18) $6x \times x^5 =$

19) $4x^2 \times 6x^6 =$

20) $5yx^3 \times 4x =$

21) $7x^3 \times y^5 x^7 =$

22) $y^2 x^3 \times y^5 x^4 =$

23) $3x^5 \times 4x^3 y^4 =$

24) $4x^4 \times 9x^2 y^5 =$

25) $5x^3 y^4 \times 6x^8 y^2 =$

26) $8x^3 y^6 \times 4xy^3 =$

27) $2xy^5 \times 6x^3 y^3 =$

28) $4x^5 y^2 \times 4x^2 y^8 =$

29) $7x \times 3y^8 x^2 \times y^5 =$

30) $x^3 \times 2y^3 x^4 \times 2y =$

31) $3yx^4 \times 3y^4 x \times 3xy^3 =$

32) $6y^3 \times 2y^2 x^4 \times 10yx^5 =$

WWW.MathNotion.Com

ISEE Middle-Level Subject Test Mathematics

Zero and Negative Exponents

✎ **Evaluate the following expressions.**

1) $1^{-5} =$

2) $4^{-1} =$

3) $0^{10} =$

4) $1^{15} =$

5) $5^{-2} =$

6) $3^{-3} =$

7) $9^{-1} =$

8) $10^{-2} =$

9) $12^{-2} =$

10) $2^{-5} =$

11) $3^{-4} =$

12) $2^{-4} =$

13) $6^{-3} =$

14) $10^{-3} =$

15) $30^{-1} =$

16) $15^{-2} =$

17) $4^{-3} =$

18) $2^{-7} =$

19) $5^{-3} =$

20) $4^{-4} =$

21) $3^{-5} =$

22) $10^{-4} =$

23) $2^{-10} =$

24) $8^{-3} =$

25) $20^{-2} =$

26) $14^{-2} =$

27) $9^{-3} =$

28) $100^{-2} =$

29) $5^{-4} =$

30) $4^{-6} =$

31) $\left(\frac{1}{4}\right)^{-3} =$

32) $\left(\frac{1}{6}\right)^{-2} =$

33) $\left(\frac{1}{7}\right)^{-2} =$

34) $\left(\frac{2}{3}\right)^{-3} =$

35) $\left(\frac{1}{13}\right)^{-2} =$

36) $\left(\frac{7}{12}\right)^{-2} =$

37) $\left(\frac{1}{6}\right)^{-3} =$

38) $\left(\frac{1}{300}\right)^{-2} =$

39) $\left(\frac{2}{9}\right)^{-2} =$

40) $\left(\frac{7}{5}\right)^{-1} =$

41) $\left(\frac{13}{23}\right)^{0} =$

42) $\left(\frac{1}{4}\right)^{-5} =$

WWW.MathNotion.Com

ISEE Middle-Level Subject Test Mathematics

Division Property of Exponents

✏️ **Simplify.**

1) $\dfrac{5^6}{5^7} =$

2) $\dfrac{8^8}{8^6} =$

3) $\dfrac{4^5}{4} =$

4) $\dfrac{3}{3^5} =$

5) $\dfrac{x}{x^6} =$

6) $\dfrac{3 \times 3^2}{3^2 \times 3^5} =$

7) $\dfrac{9^4}{9^2} =$

8) $\dfrac{10 \times 10^9}{10^2 \times 10^7} =$

9) $\dfrac{7^5 \times 7^7}{7^4 \times 7^8} =$

10) $\dfrac{15x}{30x^6} =$

11) $\dfrac{3x^9}{4x^4} =$

12) $\dfrac{15x^8}{10x^9} =$

13) $\dfrac{42x^5}{6y^9} =$

14) $\dfrac{36y^8}{4x^4y^5} =$

15) $\dfrac{2x^7}{9x} =$

16) $\dfrac{49x^8y^6}{7x^9} =$

17) $\dfrac{48x^2}{24x^6y^{12}} =$

18) $\dfrac{30yx^5}{6yx^7} =$

19) $\dfrac{19x^7y}{38x^{12}y^4} =$

20) $\dfrac{9x^8}{63x^8} =$

21) $\dfrac{9x^{-9}}{4x^{-3}} =$

WWW.MathNotion.Com

ISEE Middle-Level Subject Test Mathematics

Powers of Products and Quotients

✎ **Simplify.**

1) $(4^3)^2 =$

2) $(2^3)^4 =$

3) $(2 \times 2^3)^2 =$

4) $(5 \times 5^5)^6 =$

5) $(19^4 \times 19^2)^3 =$

6) $(2^3 \times 2^4)^4 =$

7) $(5 \times 5^2)^2 =$

8) $(4^4)^4 =$

9) $(8x^5)^2 =$

10) $(3x^2 y^4)^4 =$

11) $(7x^5 y^2)^2 =$

12) $(5x^4 y^4)^3 =$

13) $(2x^3 y^3)^5 =$

14) $(10x^3 y^4)^3 =$

15) $(13y^3 y)^2 =$

16) $(5x^6 x^4)^2 =$

17) $(6x^7 y^6)^3 =$

18) $(12x^5 x^7)^2 =$

19) $(2x^4 \times 2x)^4 =$

20) $(2x^4 y^3)^5 =$

21) $(15x^7 y^2)^2 =$

22) $(8x^3 y^5)^3 =$

23) $(3x \times 2y^2)^4 =$

24) $\left(\frac{4x}{x^5}\right)^2 =$

25) $\left(\frac{x^4 y^5}{x^3 y^5}\right)^9 =$

26) $\left(\frac{36xy}{6x^5}\right)^3 =$

27) $\left(\frac{x^7}{x^8 y^2}\right)^6 =$

28) $\left(\frac{xy^4}{x^3 y^6}\right)^{-3} =$

29) $\left(\frac{5xy^8}{x^3}\right)^2 =$

30) $\left(\frac{xy^6}{2xy^3}\right)^{-4} =$

ISEE Middle-Level Subject Test Mathematics

Negative Exponents and Negative Bases

✏️ **Simplify.**

1) $-9^{-1} =$

2) $-9^{-2} =$

3) $-2^{-5} =$

4) $-x^{-7} =$

5) $11x^{-1} =$

6) $-8x^{-3} =$

7) $-12x^{-5} =$

8) $-9x^{-8}y^{-6} =$

9) $32x^{-5}y^{-1} =$

10) $10a^{-9}b^{-3} =$

11) $-17x^4y^{-6} =$

12) $-\frac{25}{x^{-5}} =$

13) $-\frac{13x}{a^{-7}} =$

14) $(-\frac{1}{3})^{-4} =$

15) $(-\frac{3}{4})^{-2} =$

16) $-\frac{14}{a^{-6}b^{-3}} =$

17) $-\frac{7x}{x^{-8}} =$

18) $-\frac{a^{-9}}{b^{-5}} =$

19) $-\frac{11}{x^{-5}} =$

20) $\frac{8b}{-16c^{-6}} =$

21) $\frac{12ab}{a^{-4}b^{-3}} =$

22) $-\frac{8n^{-4}}{32p^{-7}} =$

23) $\frac{16ab^{-6}}{-6c^{-5}} =$

24) $(\frac{10a}{5c})^{-4} =$

25) $(-\frac{12x}{4yz})^{-3} =$

26) $\frac{8ab^{-7}}{-5c^{-3}} =$

27) $(-\frac{x^4}{x^5})^{-5} =$

28) $(-\frac{x^{-2}}{7x^3})^{-2} =$

29) $(-\frac{x^{-4}}{x^2})^{-6} =$

WWW.MathNotion.Com

ISEE Middle-Level Subject Test Mathematics

Scientific Notation

✎ Write each number in scientific notation.

1) $0.223 =$

2) $0.09 =$

3) $4.5 =$

4) $900 =$

5) $2,000 =$

6) $0.006 =$

7) $33 =$

8) $9,400 =$

9) $1,470 =$

10) $52,000 =$

11) $8,000,000 =$

12) $0.00009 =$

13) $2,158,000 =$

14) $0.0039 =$

15) $0.000075 =$

16) $4,300,000 =$

17) $130,000 =$

18) $4,000,000,000 =$

19) $0.00009 =$

20) $0.0039 =$

✎ Write each number in standard notation.

21) $4 \times 10^{-1} =$

22) $1.2 \times 10^{-3} =$

23) $2.7 \times 10^{5} =$

24) $6 \times 10^{-4} =$

25) $3.6 \times 10^{-3} =$

26) $5.5 \times 10^{5} =$

27) $3.2 \times 10^{4} =$

28) $3.88 \times 10^{6} =$

29) $7 \times 10^{-6} =$

30) $4.2 \times 10^{-7} =$

WWW.MathNotion.Com

ISEE Middle-Level Subject Test Mathematics

Square Roots

✎ Find the value each square root.

1) $\sqrt{16} =$ ___
2) $\sqrt{25} =$ ___
3) $\sqrt{1} =$ ___
4) $\sqrt{64} =$ ___
5) $\sqrt{0} =$ ___
6) $\sqrt{196} =$ ___
7) $\sqrt{4} =$ ___
8) $\sqrt{256} =$ ___

9) $\sqrt{36} =$ ___
10) $\sqrt{289} =$ ___
11) $\sqrt{169} =$ ___
12) $\sqrt{144} =$ ___
13) $\sqrt{100} =$ ___
14) $\sqrt{1,600} =$ ___
15) $\sqrt{2,500} =$ ___
16) $\sqrt{324} =$ ___

17) $\sqrt{529} =$ ___
18) $\sqrt{20} =$ ___
19) $\sqrt{625} =$ ___
20) $\sqrt{18} =$ ___
21) $\sqrt{50} =$ ___
22) $\sqrt{1,024} =$ ___
23) $\sqrt{160} =$ ___
24) $\sqrt{32} =$ ___

✎ Evaluate.

25) $\sqrt{4} \times \sqrt{25} =$ _____
26) $\sqrt{36} \times \sqrt{49} =$ _____
27) $\sqrt{6} \times \sqrt{6} =$ _____
28) $\sqrt{13} \times \sqrt{13} =$ _____
29) $2\sqrt{5} \times 3\sqrt{5} =$ _____
30) $\sqrt{12} \times \sqrt{3} =$ _____

31) $\sqrt{13} + \sqrt{13} =$ _____
32) $\sqrt{10} + 2\sqrt{10} =$ _____
33) $12\sqrt{7} - 10\sqrt{7} =$ _____
34) $4\sqrt{10} \times 2\sqrt{10} =$ _____
35) $5\sqrt{3} \times 8\sqrt{3} =$ _____
36) $6\sqrt{3} - \sqrt{12} =$ _____

WWW.MathNotion.Com

ISEE Middle-Level Subject Test Mathematics

Answers of Worksheets

Add and Subtract Exponents.

1) 41
2) $2x^6$
3) b^2
4) 67
5) −8
6) 11
7) $5x^3$
8) 343
9) 1,008
10) 24
11) 0
12) 15
13) −119
14) 54
15) 17
16) 0
17) −5
18) 336
19) 28
20) 25
21) 72
22) 37
23) $14x^5$
24) 17
25) 18
26) 125
27) $\frac{1}{2}$
28) 90

Multiplication Property of Exponents

1) 4^6
2) 8^5
3) 7^6
4) 9^4
5) 2^7
6) 5^7
7) 4^7
8) $5x^2$
9) x^6
10) x^9
11) x^9
12) $30x^2$
13) $16x^6$
14) $7x^4$
15) $12x^6$
16) $5x^8$
17) $4x^9$
18) $6x^6$
19) $24x^8$
20) $20x^4y$
21) $7x^{10}y^5$
22) x^7y^7
23) $12x^8y^4$
24) $36x^6y^5$
25) $30x^{11}y^6$
26) $32x^4y^9$
27) $12x^4y^8$
28) $16x^7y^{10}$
29) $21x^3y^{13}$
30) $4x^7y^4$
31) $27x^6y^8$
32) $120x^9y^6$

Zero and Negative Exponents

1) 1
2) $\frac{1}{4}$
3) 0
4) 1
5) $\frac{1}{25}$
6) $\frac{1}{27}$
7) $\frac{1}{9}$
8) $\frac{1}{100}$
9) $\frac{1}{144}$
10) $\frac{1}{32}$
11) $\frac{1}{81}$
12) $\frac{1}{16}$
13) $\frac{1}{216}$
14) $\frac{1}{1,000}$
15) $\frac{1}{30}$
16) $\frac{1}{225}$
17) $\frac{1}{64}$
18) $\frac{1}{128}$
19) $\frac{1}{125}$
20) $\frac{1}{256}$
21) $\frac{1}{243}$
22) $\frac{1}{10,000}$
23) $\frac{1}{1,024}$
24) $\frac{1}{512}$
25) $\frac{1}{400}$

WWW.MathNotion.Com

ISEE Middle-Level Subject Test Mathematics

26) $\frac{1}{196}$
27) $\frac{1}{729}$
28) $\frac{1}{10,000}$
29) $\frac{1}{625}$
30) $\frac{1}{4,096}$
31) 64
32) 36
33) 49
34) $\frac{27}{8}$
35) 169
36) $\frac{144}{49}$
37) 216
38) 90,000
39) $\frac{81}{4}$
40) $\frac{5}{7}$
41) 1
42) 1,024

Division Property of Exponents

1) $\frac{1}{5}$
2) 8^2
3) 4^4
4) $\frac{1}{3^4}$
5) $\frac{1}{x^5}$
6) $\frac{1}{3^4}$
7) 9^2
8) 10
9) 1
10) $\frac{1}{2x^5}$
11) $\frac{3x^5}{4}$
12) $\frac{3}{2x}$
13) $\frac{7x^5}{y^9}$
14) $\frac{9y^3}{x^4}$
15) $\frac{2x^6}{9}$
16) $\frac{7y^6}{x}$
17) $\frac{2}{x^4 y^{12}}$
18) $\frac{5}{x^2}$
19) $\frac{1}{2x^5 y^3}$
20) $\frac{1}{7}$
21) $\frac{9}{4x^6}$

Powers of Products and Quotients

1) 4^6
2) 2^{12}
3) 2^8
4) 5^{36}
5) 19^{18}
6) 2^{28}
7) 5^6
8) 4^{16}
9) $64x^{10}$
10) $81x^8 y^{16}$
11) $49x^{10} y^4$
12) $125x^{12} y^{12}$
13) $32x^{15} y^{15}$
14) $1,000x^9 y^{12}$
15) $169y^8$
16) $25x^{20}$
17) $216x^{21} y^{18}$
18) $144x^{24}$
19) $256x^{20}$
20) $32x^{20} y^{15}$
21) $225x^{14} y^4$
22) $512x^9 y^{15}$
23) $1,296x^4 y^8$
24) $\frac{16}{x^8}$
25) x^9
26) $\frac{216y^3}{x^{12}}$
27) $\frac{1}{x^6 y^{12}}$
28) $x^6 y^6$
29) $\frac{25y^{16}}{x^4}$
30) $\frac{16}{y^{12}}$

Negative Exponents and Negative Bases

1) $-\frac{1}{9}$
2) $-\frac{1}{81}$
3) $-\frac{1}{32}$
4) $-\frac{1}{x^7}$
5) $\frac{11}{x}$
6) $-\frac{8}{x^3}$
7) $-\frac{12}{x^5}$
8) $-\frac{9}{x^8 y^6}$
9) $\frac{32}{x^5 y}$

ISEE Middle-Level Subject Test Mathematics

10) $\frac{10}{a^9 b^3}$

11) $-\frac{17x^4}{y^6}$

12) $-25x^5$

13) $-13xa^7$

14) 81

15) $\frac{16}{9}$

16) $-14a^6 b^3$

17) $-7x^9$

18) $-\frac{b^5}{a^9}$

19) $-11x^5$

20) $-\frac{bc^6}{2}$

21) $12a^5 b^4$

22) $-\frac{p^7}{4n^4}$

23) $-\frac{8ac^5}{3b^6}$

24) $\frac{c^4}{16a^4}$

25) $\frac{y^3 z^3}{27 x^3}$

26) $-\frac{8ac^3}{5b^7}$

27) $-x^5$

28) $49x^{10}$

29) x^{36}

Scientific Notation

1) 2.23×10^{-1}
2) 9×10^{-2}
3) 4.5×10^0
4) 9×10^2
5) 2×10^3
6) 6×10^{-3}
7) 3.3×10^1
8) 9.4×10^3
9) 1.47×10^3
10) 5.2×10^4
11) 8×10^6
12) 9×10^{-5}
13) 2.158×10^6
14) 3.9×10^{-3}
15) 7.5×10^{-5}
16) 4.3×10^6
17) 1.3×10^5
18) 4×10^9
19) 9×10^{-5}
20) 3.9×10^{-3}
21) 0.4
22) 0.0012
23) $270{,}000$
24) 0.0006
25) 0.0036
26) $550{,}000$
27) $32{,}000$
28) $3{,}880{,}000$
29) 0.000007
30) 0.00000042

Square Roots

1) 4
2) 5
3) 1
4) 8
5) 0
6) 14
7) 2
8) 16
9) 6
10) 17
11) 13
12) 12
13) 10
14) 40
15) 50
16) 18
17) 23
18) $2\sqrt{5}$
19) 25
20) $3\sqrt{2}$
21) $5\sqrt{2}$
22) 32
23) $4\sqrt{10}$
24) $4\sqrt{2}$
25) 10
26) 42
27) 6
28) 13
29) 30
30) 6
31) $2\sqrt{13}$
32) $3\sqrt{10}$
33) $2\sqrt{7}$
34) 80
35) 120
36) $4\sqrt{3}$

WWW.MathNotion.Com

ISEE Middle-Level Subject Test Mathematics

Chapter 7 : Measurements

Topics that you will learn in this chapter:

- ✓ Reference Measurement
- ✓ Metric Length
- ✓ Customary Length
- ✓ Metric Capacity
- ✓ Customary Capacity
- ✓ Metric Weight and Mass
- ✓ Customary Weight and Mass
- ✓ Temperature
- ✓ Time

"It's not that I'm so smart, it's just that I stay with problems longer." -Albert Einstein

ISEE Middle-Level Subject Test Mathematics

Reference Measurement

LENGTH	
Customary	**Metric**
1 mile (mi) = 1,760 yards (yd)	1 kilometer (km) = 1,000 meters (m)
1 yard (yd) = 3 feet (ft)	1 meter (m) = 100 centimeters (cm)
1 foot (ft) = 12 inches (in.)	1 centimeter(cm) = 10 millimeters(mm)
VOLUME AND CAPACITY	
Customary	**Metric**
1 gallon (gal) = 4 quarts (qt)	1 liter (L) = 1,000 milliliters (mL)
1 quart (qt) = 2 pints (pt.)	
1 pint (pt.) = 2 cups (c)	
1 cup (c) = 8 fluid ounces (Fl oz)	
WEIGHT AND MASS	
Customary	**Metric**
1 ton (T) = 2,000 pounds (lb.)	1 kilogram (kg) = 1,000 grams (g)
1 pound (lb.) = 16 ounces (oz)	1 gram (g) = 1,000 milligrams (mg)
Time	
1 year = 12 months	
1 year = 52 weeks	
1 week = 7 days	
1 day = 24 hours	
1 hour = 60 minutes	
1 minute = 60 seconds	

ISEE Middle-Level Subject Test Mathematics

Metric Length Measurement

✎ **Convert to the units.**

1) 3×10^3 mm = _____ cm

2) 0.95 m = _____ mm

3) 0.08 m = _____ cm

4) 2.25 km = _____ m

5) 7,800 mm = _____ m

6) 9,100 cm = _____ m

7) 5.83 m = _____ cm

8) 2×10^5 mm = _____ cm

9) 8×10^3 mm = _____ m

10) 0.003 km = _____ mm

11) 0.7 km = _____ m

12) 0.011 m = _____ cm

13) 125×10^5 m = _____ km

14) 78×10^4 m = _____ km

Customary Length Measurement

✎ **Convert to the units.**

1) 15 ft = _____ in

2) 1.5 ft = _____ in

3) 4.8 yd = _____ ft

4) 0.82 yd = _____ ft

5) 17×10^{-3} yd = _____ in

6) 0.5 mi = _____ in

7) 1,746 in = _____ yd

8) 3.24 in = _____ yd

9) 3,960 yd = _____ mi

10) 42.55 yd = _____ in

11) 5×10^{-2} mi = _____ yd

12) 87,120 ft = _____ mi

13) 2.52 in = _____ ft

14) 29.3 yd = _____ feet

15) 0.612 in = _____ ft

16) 1.3 mi = _____ ft

WWW.MathNotion.Com

ISEE Middle-Level Subject Test Mathematics

Metric Capacity Measurement

✎ **Convert the following measurements.**

1) 1.58 l = _____ ml
2) 0.504 l = _____ ml
3) 3.04 l = _____ ml
4) 0.005 l = _____ ml
5) 121.56 l = _____ ml
6) 0.0459 l = _____ ml
7) 4.2×10^5 ml = _____ l
8) 3.12×10^3 ml = _____ l
9) $1,889 \times 10^2$ ml = _____ l
10) 250 ml = _____ l
11) 656,160 ml = _____ l
12) 0.54×10^4 ml = _____ l

Customary Capacity Measurement

✎ **Convert the following measurements.**

1) 0.7 gal = _____ qt.
2) 3.2 gal = _____ pt.
3) 0.75 gal = _____ c.
4) 15.5 pt. = _____ c
5) 18.2 c = _____ fl oz
6) 9.02 qt = _____ pt.
7) 1.05 qt = _____ c
8) 158 pt. = _____ c
9) 9.6×10^3 c = _____ gal
10) 203.2 pt. = _____ gal
11) 12.4 qt = _____ gal
12) 115.6 pt. = _____ qt
13) 4,880 c = _____ qt
14) 113.6 c = _____ pt.
15) 0.036 qt = _____ gal
16) 522.4 pt. = _____ qt
17) 5.8 gal = _____ pt.
18) 0.002 qt = _____ c
19) 672 c = _____ gal
20) 72.96 fl oz = _____ c

WWW.MathNotion.Com

ISEE Middle-Level Subject Test Mathematics

Metric Weight and Mass Measurement

✏️ **Convert.**

1) 0.712 kg = _____ g

2) 54.01 kg = _____ g

3) 9.8×10^{-5} kg = _____ g

4) 0.012 kg = _____ g

5) 120.02 kg = _____ g

6) 1.199 kg = _____ g

7) 0.0055 kg = _____ g

8) 9×10^4 g = _____ kg

9) 3.5×10^5 g = _____ kg

10) 0.008×10^4 g = _____ kg

11) 15,010 g = _____ kg

12) 12.1×10^4 g = _____ kg

13) 4,155,200 g = _____ kg

14) 402×10^2 g = _____ kg

Customary Weight and Mass Measurement

✏️ **Convert.**

1) 36×10^2 lb. = _____ T

2) 0.022×10^4 lb. = _____ T

3) 215,000 lb. = _____ T

4) 12,600 lb. = _____ T

5) 0.015 lb. = _____ oz

6) 1.6 lb. = _____ oz

7) 0.021 lb. = _____ oz

8) 5.2 T = _____ lb.

9) 6.8×10^{-3} T = _____ lb.

10) 156×10^{-2} T = _____ lb.

11) 0.017 T = _____ lb.

12) 1.085 T = _____ oz

13) 0.006 T = _____ oz

14) 209.92 oz = _____ lb.

WWW.MathNotion.Com

ISEE Middle-Level Subject Test Mathematics

Temperature

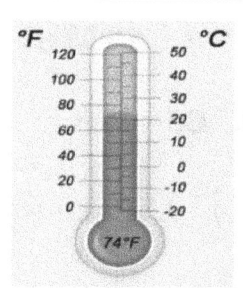

✎ **Convert Fahrenheit into Celsius.**

1) 35.6°F = ___ °C

2) 54.5°F = ___ °C

3) −25.6°F = ___ °C

4) 62.6°F = ___ °C

5) 120.2°F = ___ °C

6) 174.2°F = ___ °C

7) 51.8°F = ___ °C

8) 193.1°F = ___ °C

9) 221°F = ___ °C

10) 60.44°F = ___ °C

11) 48.2°F = ___ °C

12) 134.6°F = ___ °C

✎ **Convert Celsius into Fahrenheit.**

13) 18.2°C = ___ °F

14) 88.8°C = ___ °F

15) 250°C = ___ °F

16) 52°C = ___ °F

17) 10°C = ___ °F

18) −24°C = ___ °F

19) 6°C = ___ °F

20) 15.6°C = ___ °F

21) 33°C = ___ °F

22) 61°C = ___ °F

23) 113.5°C = ___ °F

24) 28°C = ___ °F

WWW.MathNotion.Com

ISEE Middle-Level Subject Test Mathematics

Time

✎ **Convert to the units.**

1) 18.5 hr. = _____ min

2) 26 year = _____ week

3) 0.2 hr. = _____ sec

4) 12.5 min = _____ sec

5) 1.2×10^5 min = _____ hr

6) 1,460 day = _____ year

7) 3 year = _____ hr.

8) 51 day = _____ hr

9) 5 day = _____ min

10) 552 min = _____ hr

11) 32.25 year = _____ month

12) 6,480 sec = _____ min

13) 264 hr = _____ day

14) 22 weeks = _____ day

✎ **How much time has passed?**

1) From 1:45 A.M. to 4:55 A.M.: ____ hours and ____ minutes.

2) From 1:25 A.M. to 6:05 A.M.: ____ hours and ____ minutes.

3) It's 7:15 P.M. What time was 4 hours ago? _____ O'clock

4) 3:05 A.M to 6:55 AM: ____ hours and ____ minutes.

5) 3:45 A.M to 5:15 AM: ____ hours and ____ minutes.

6) 8:05 A.M. to 11:20 AM. = ____ hour(s) and ____ minutes.

7) 10:55 A.M. to 1:25 PM. = ____ hour(s) and ____ minutes

8) 6:18 A.M. to 6:52 A.M. = _____ minutes

9) 3:54 A.M. to 4:08 A.M. = _____ minutes

WWW.MathNotion.Com

ISEE Middle-Level Subject Test Mathematics

Answers of Worksheets

Metric length

1) 300 cm
2) 950 mm
3) 8 cm
4) 2,250 m
5) 7.8 m
6) 91 m
7) 583 cm
8) 20,000 cm
9) 8 m
10) 3,000 mm
11) 700 m
12) 1.1 cm
13) 12,500 km
14) 780km

Customary Length

1) 180
2) 18
3) 14.4
4) 2.46
5) 0.612
6) 31,680
7) 48.5
8) 0.09
9) 2.25
10) 1,531.8
11) 88
12) 16.5
13) 0.21
14) 87.9
15) 0.051
16) 6,864

Metric Capacity

1) 1,580 ml
2) 504 ml
3) 3,040 ml
4) 5 ml
5) 121,560 ml
6) 45.9 ml
7) 420 L
8) 3.12 L
9) 188.9 L
10) 0.25L
11) 656.16 L
12) 5.4 L

Customary Capacity

1) 2.8 qt
2) 25.6 pt.
3) 12 c
4) 31 c
5) 145.6 fl oz
6) 18.04 pt.
7) 4.2 c
8) 316 c
9) 600 gal
10) 25.4 gal
11) 3.1 gal
12) 57.8 qt
13) 1,220qt
14) 56.8 pt.
15) 0.009 gal
16) 261.2 qt
17) 46.4 pt.
18) 0.008 c
19) 42 gal
20) 9.12 c

Metric Weight and Mass

1) 712 g
2) 54,010 g
3) 0.098 g
4) 12 g
5) 120,020 g
6) 1,199 g
7) 5.5 g
8) 90 kg
9) 350 kg
10) 0.08 kg
11) 15.01 kg
12) 121 kg
13) 4,155.2 kg
14) 40.2 kg

WWW.MathNotion.Com

ISEE Middle-Level Subject Test Mathematics

Customary Weight and Mass

1) 1.8 T
2) 0.11 T
3) 107.5 T
4) 6.3 T
5) 0.24 oz
6) 25.6 oz
7) 0.336 oz
8) 10,400 lb.
9) 13.6 lb.
10) 3,120 lb.
11) 34 lb.
12) 34,720 oz
13) 192 oz
14) 13.12 lb

Temperature

1) 2°C
2) 12.5°C
3) −32°C
4) 17°C
5) 49°C
6) 79°C
7) 11°C
8) 89.5°C
9) 105°C
10) 15.8°C
11) 9°C
12) 57°C
13) 64.76°F
14) 191.84°F
15) 482°F
16) 125.6°F
17) 50°F
18) −11.2°F
19) 42.8°F
20) 60.08°F
21) 91.4°F
22) 141.8°F
23) 236.3°F
24) 82.4°F

Time - Convert.

1) 1,110 min
2) 1,352 weeks
3) 720 sec
4) 750 sec
5) 2,000 hr
6) 4 year
7) 26,280 hr
8) 1,224 hr
9) 7,200 min
10) 9.2 hr
11) 387 months
12) 108 min
13) 11 days
14) 154 days

Time - Gap

1) 3:10
2) 4:40
3) 3:15 P.M.
4) 3:50
5) 1:30
6) 3:15
7) 2:30
8) 34 minutes
9) 14 minutes

WWW.MathNotion.Com

ISEE Middle-Level Subject Test Mathematics

Chapter 8:
Algebraic Expressions

Topics that you will practice in this chapter:

- ✓ Find a rule!
- ✓ Translate Phrases into an Algebraic Statement
- ✓ Simplifying Variable Expressions
- ✓ The Distributive Property
- ✓ Evaluating One Variable Expressions
- ✓ Evaluating Two Variables Expressions
- ✓ Combining like Terms

Mathematics is, as it were, a sensuous logic, and relates to philosophy as do the arts, music, and plastic art to poetry. — *K. Shegel*

Find a Rule!

✍ **Complete the output.**

1- **Rule:** the output is $x - 10.5$

Input	x	15	18	27	32.25	48.5
Output	y					

1) **Rule:** the output is $x \times 5\frac{1}{3}$

Input	x	3	9	15	21	33
Output	y					

2- **Rule:** the output is $x \div 9$

Input	x	513	387	342	198	126
Output	y					

✍ **Find a rule to write an expression.**

3- **Rule:** _____

Input	x	4	14	19	24
Output	y	10	35	47.5	60

4- **Rule:** _____

Input	x	5	13	19.6	34.5
Output	y	14.4	22.4	29	43.9

5- **Rule:** _____

Input	x	72	96	132	230.4
Output	y	9	12	16.5	28.8

ISEE Middle-Level Subject Test Mathematics

Translate Phrases into an Algebraic Statement

✏ **Write an algebraic expression for each phrase.**

1) 9 multiplied by x. _____

2) Subtract 11 from y. _____

3) 19 divided by x. _____

4) 38 decreased by y. _____

5) Add y to 40. _____

6) The square of 6. _____

7) x raised to the fifth power. _____

8) The sum of six and a number. _____

9) The difference between fifty-seven and y. _____

10) The quotient of nine and a number. _____

11) The quotient of the square of x and 25. _____

12) The difference between x and 6 is 19. _____

13) 10 times a reduced by the square of b. _____

14) Subtract the product of a and b from 41. _____

ISEE Middle-Level Subject Test Mathematics

Simplifying Variable Expressions

✍ **Simplify each expression.**

1) $3(x + 5) =$

2) $(-4)(7x - 5) =$

3) $11x + 5 - 6x =$

4) $-4 - 2x^2 - 6x^2 =$

5) $7 + 13x^2 + 3 =$

6) $3x^2 + 7x + 15x^2 =$

7) $3x^2 - 12x^2 + 4x =$

8) $4x^2 - 8x - 2x =$

9) $6x + 7(3 - 4x) =$

10) $8x + 4(15x - 3) =$

11) $6(-3x - 9) - 17 =$

12) $-11x^2 - (-5x) =$

13) $2x + 7 + 5 - 8x =$

14) $7 + 6x - 11 - 5x =$

15) $27x + 8 - 13 - 5x =$

16) $(-11)(-5x + 2) - 41x =$

17) $19x - 4(4 - 2x) =$

18) $16x + 3(3x + 6) + 10 =$

19) $5(-2x - 4) - 13x =$

20) $16x - 3x(x + 10) =$

21) $17x + 5x(2 - 4x) =$

22) $5x(-4x - 7) + 20x =$

23) $25x - 19 + 4x^2 =$

24) $6x(x - 11) + 25 =$

25) $4x - 5 + 15x + 3x^2 =$

26) $-7x^2 - 11x - 9x =$

27) $10x - 9x^2 - 3x^2 - 7 =$

28) $13 + 3x^2 - 9x^2 - 21x =$

29) $22x + 10x^2 - 15x + 17 =$

30) $4x^2 + 25x + 21x^2 =$

31) $29 - 12x^2 - 23x - 4x^2 =$

32) $22x - 19x - 9x^2 + 30 =$

WWW.MathNotion.Com

ISEE Middle-Level Subject Test Mathematics

The Distributive Property

✏️ Use the distributive property to simply each expression.

1) $4(1 + 2x) =$

2) $2(4 + 7x) =$

3) $3(4x - 4) =$

4) $(2x - 5)(-6) =$

5) $(-3)(x + 6) =$

6) $(4 + 3x)2 =$

7) $(-5)(8 - 3x) =$

8) $-(-5 - 7x) =$

9) $(-6x + 3)(-3) =$

10) $(-4)(x - 7) =$

11) $-(5 - 3x) =$

12) $3(9 + 4x) =$

13) $6(4 + 3x) =$

14) $(-5x + 3)2 =$

15) $(5 - 8x)(-3) =$

16) $(-12)(3x + 3) =$

17) $(5 - 3x)6 =$

18) $4(2 + 6x) =$

19) $8(7x - 3) =$

20) $(-2x + 3)4 =$

21) $(7 - 5x)(-9) =$

22) $(-10)(x - 8) =$

23) $(11 - 4x)3 =$

24) $(-6)(10x - 4) =$

25) $(3 - 9x)(-7) =$

26) $(-9)(x + 9) =$

27) $(-3 + 5x)(-7) =$

28) $(-5)(8 - 10x) =$

29) $12(4x - 8) =$

30) $(-10x + 13)(-3) =$

31) $(-8)(3x - 2) + 4(x + 5) =$

32) $(-8)(x + 4) - (6 + 5x) =$

WWW.MathNotion.Com

ISEE Middle-Level Subject Test Mathematics

Evaluating One Variable Expressions

✎ **Evaluate each expression using the value given.**

1) $8 - x, x = 5$

2) $x - 9, x = 5$

3) $5x + 4, x = 3$

4) $x - 13, x = -4$

5) $12 - x, x = 4$

6) $x + 2, x = 6$

7) $4x + 8, x = 3$

8) $x + (-7), x = -8$

9) $4x + 5, x = 2$

10) $3x + 9, x = -2$

11) $15 + 3x - 7, x = 2$

12) $17 - 3x, x = 3$

13) $8x - 9, x = 4$

14) $5x + 4, x = -3$

15) $10x + 5, x = 3$

16) $14 - 4x, x = -6$

17) $3(5x + 3), x = 9$

18) $4(-3x - 6), x = 3$

19) $7x - 2x + 12, x = 4$

20) $(5x + 6) \div 2, x = 8$

21) $(x + 18) \div 10, x = 12$

22) $5x - 12 + 3x, x = -3$

23) $(6 - 4x)(-3), x = -4$

24) $9x^2 + 3x - 6, x = 2$

25) $x^2 - 10x, x = -5$

26) $3x(7 - 2x), x = 2$

27) $12x + 6 - 2x^2, x = -4$

28) $(-3)(4x - 8 + 3x), x = 3$

29) $(-6) + \frac{x}{4} + 3x, x = 16$

30) $(-6) + \frac{x}{5}, x = 35$

31) $\left(-\frac{45}{x}\right) - 7 + 2x, x = 9$

32) $\left(-\frac{21}{x}\right) - 12 + 4x, x = 7$

WWW.MathNotion.Com

ISEE Middle-Level Subject Test Mathematics

Combining like Terms

✏️ **Simplify each expression.**

1) $11x + 3x + 6 =$

2) $8(2x - 6) =$

3) $18x - 7x + 11 =$

4) $(-4)(6x - 7) =$

5) $22x - 10x - 5 =$

6) $32x - 13 + 8x =$

7) $15 - (8x - 11) =$

8) $-24x + 17 - 11x =$

9) $12x - 8 - 6x + 9 =$

10) $21x + 5 - 36 + 12x =$

11) $28x + 3x - 11 =$

12) $(-3x + 4)5 =$

13) $2 + 4x + 9x - 8 =$

14) $6(2x - 5x) - 4 =$

15) $4(5x + 11) + 3x =$

16) $x - 14 - 11x =$

17) $5(10 + 9x) - 8x =$

18) $42x + 17 - 23x =$

19) $(-7x) + 19 + 20x =$

20) $(-7x) - 33 + 29x =$

21) $4(5x + 3) - 19x =$

22) $5(6 - 2x) - 15x =$

23) $-24x + (11 - 18x) =$

24) $(-9) - (6)(7x + 3) =$

25) $(-1)(8x - 10) - 21x =$

26) $-36x + 14 + 27x - 5x =$

27) $3(-13x + 6) - 17x =$

28) $-5x - 42 + 32x =$

29) $37x - 19x + 15 - 9x =$

30) $3(5x + 7x) - 31 =$

31) $14 - 6x - 15 - 9x =$

32) $-2(-5x - 7x) + 27x =$

WWW.MathNotion.Com

ISEE Middle-Level Subject Test Mathematics

Answers of Worksheets

Find a rule.

1)
Input	x	15	18	27	32.25	48.5
Output	y	4.5	7.5	16.5	21.75	38

2)
Input	x	3	9	15	21	33
Output	y	16	48	80	112	176

3)
Input	x	513	387	342	198	126
Output	y	57	43	38	22	14

4) $y = 2.5x$ 5) $y = x + 9.4$ 6) $y = x \div 8$

Translate Phrases into an Algebraic Statement

1) $9x$
2) $y - 11$
3) $\frac{19}{x}$
4) $38 - y$
5) $y + 40$
6) 6^2
7) x^5
8) $6 + x$
9) $57 - y$
10) $\frac{9}{x}$
11) $\frac{x^2}{25}$
12) $x - 6 = 19$
13) $10a - b^2$
14) $41 - ab$

Simplifying Variable Expressions

1) $3x + 15$
2) $-28x + 20$
3) $5x + 5$
4) $-8x^2 - 4$
5) $13x^2 + 10$
6) $18x^2 + 7x$
7) $-9x^2 + 4x$
8) $4x^2 - 10x$
9) $-22x + 21$
10) $68x - 12$
11) $-18x - 71$
12) $-11x^2 + 5x$
13) $-6x + 12$
14) $x - 4$
15) $22x - 5$
16) $14x - 22$
17) $27x - 16$
18) $25x + 28$
19) $-23x - 20$
20) $-3x^2 - 14x$
21) $-20x^2 + 27x$
22) $-20x^2 - 15x$
23) $4x^2 + 25x - 19$
24) $6x^2 - 66x + 25$
25) $3x^2 + 19x - 5$
26) $-7x^2 - 20x$
27) $-12x^2 + 10x - 7$
28) $-6x^2 - 21x + 13$
29) $10x^2 + 7x + 17$
30) $25x^2 + 25x$
31) $-16x^2 - 23x + 29$
32) $-9x^2 + 3x + 30$

The Distributive Property

1) $8x + 4$
2) $14x + 8$
3) $12x - 12$
4) $-12x + 30$
5) $-3x - 18$
6) $6x + 8$
7) $15x - 40$
8) $7x + 5$

WWW.MathNotion.Com

ISEE Middle-Level Subject Test Mathematics

9) $18x - 9$
10) $-4x + 28$
11) $3x - 5$
12) $12x + 27$
13) $18x + 24$
14) $-10x + 6$

15) $24x - 15$
16) $-36x - 36$
17) $-18x + 30$
18) $24x + 8$
19) $56x - 24$
20) $-8x + 12$

21) $45x - 63$
22) $-10x + 80$
23) $-12x + 33$
24) $-60x + 24$
25) $63x - 21$
26) $-9x - 81$

27) $-35x + 21$
28) $50x - 40$
29) $48x - 96$
30) $30x - 39$
31) $-20x + 36$
32) $-13x - 38$

Evaluating One Variables

1) 3
2) -4
3) 19
4) -17
5) 8
6) 8
7) 20
8) -15

9) 13
10) 3
11) 14
12) 8
13) 23
14) -11
15) 35
16) 38

17) 144
18) -60
19) 32
20) 23
21) 3
22) -36
23) -66
24) 36

25) 75
26) 18
27) -74
28) -39
29) 46
30) 1
31) 6
32) 13

Combining like Terms

1) $14x + 6$
2) $16x - 48$
3) $11x + 11$
4) $-24x + 28$
5) $12x - 5$
6) $40x - 13$
7) $-8x + 26$
8) $-35x + 17$

9) $6x + 1$
10) $33x - 31$
11) $31x - 11$
12) $-15x + 20$
13) $13x - 6$
14) $-18x - 4$
15) $23x + 44$
16) $-10x - 14$

17) $37x + 50$
18) $19x + 17$
19) $13x + 19$
20) $22x - 33$
21) $x + 12$
22) $-25x + 30$
23) $-42x + 11$
24) $-42x - 27$

25) $-29x + 10$
26) $-14x + 14$
27) $-56x + 18$
28) $27x - 42$
29) $9x + 15$
30) $36x - 31$
31) $-15x - 1$
32) $51x$

ISEE Middle-Level Subject Test Mathematics

Chapter 9 :
Equations and Inequalities

Topics that you will practice in this chapter:

- ✓ One–Step Equations
- ✓ Two–Step Equations
- ✓ Multi–Step Equations
- ✓ Graphing Single–Variable Inequalities
- ✓ One–Step Inequalities
- ✓ Two–Step Inequalities
- ✓ Multi-Step Inequalities

"Life is a math equation. In order to gain the most, you have to know how to convert negatives into positives." – Anonymous

ISEE Middle-Level Subject Test Mathematics

One–Step Equations

✏️ **Find the answer for each equation.**

1) $3x = 90, x = $ ____

2) $5x = 35, x = $ ____

3) $6x = 24, x = $ ____

4) $24x = 144, x = $ ____

5) $x + 15 = 20, x = $ ____

6) $x - 7 = 4, x = $ ____

7) $x - 9 = 2, x = $ ____

8) $x + 15 = 23, x = $ ____

9) $x - 4 = 13, x = $ ____

10) $12 = 16 + x, x = $ ____

11) $x - 10 = 2, x = $ ____

12) $5 - x = -11, x = $ ____

13) $28 = -6 + x, x = $ ____

14) $x - 20 = -35, x = $ ____

15) $x + 14 = -4, x = $ ____

16) $14 = 28 - x, x = $ ____

17) $7 + x = -7, x = $ ____

18) $x - 16 = 4, x = $ ____

19) $30 = x - 15, x = $ ____

20) $x - 5 = -18, x = $ ____

21) $x - 10 = 24, x = $ ____

22) $x - 20 = -25, x = $ ____

23) $x - 17 = 30, x = $ ____

24) $-70 = x - 28, x = $ ____

25) $x - 9 = 13, x = $ ____

26) $36 = 4x, x = $ ____

27) $x - 35 = 25, x = $ ____

28) $x - 25 = 10, x = $ ____

29) $70 - x = 16, x = $ ____

30) $x - 10 = 14, x = $ ____

31) $17 - x = -13, x = $ ____

32) $x - 9 = -30, x = $ ____

WWW.MathNotion.Com

ISEE Middle-Level Subject Test Mathematics

One–Step Equation Word Problems

✎ Solve.

1) How many boxes of envelopes can you buy with $40 if one box costs $5?

2) After paying $8.15 for a salad, Riya has $61.53. How much money did she have before buying the salad?

3) How many packages of Tissues can you buy with $81 if one package costs $4.5?

4) Last week Joe ran 40 miles more than Harrison. Joe ran 78 miles. How many miles did Harrison run?

5) Last Friday Liam had $65.46. Over the weekend he received some money for cleaning the attic. He now has $60. How much money did he receive?

6) After paying $17.36 for a sandwich, Elise has $31.23. How much money did she have before buying the sandwich?

ISEE Middle-Level Subject Test Mathematics

Two-Steps Equations

✎ **Solve each equation.**

1) $6(3 + x) = 42$

2) $(-7)(x - 2) = 56$

3) $(-8)(3x - 4) = (-16)$

4) $5(2 + x) = -15$

5) $19(3x + 11) = 38$

6) $4(2x + 2) = 24$

7) $5(8 + 3x) = (-20)$

8) $(-5)(5x - 3) = 40$

9) $2x + 12 = 16$

10) $\frac{4x - 5}{5} = 3$

11) $(-3) = \frac{x + 4}{7}$

12) $80 = (-8)(x - 3)$

13) $\frac{x}{3} + 7 = 19$

14) $\frac{1}{4} = \frac{1}{2} + \frac{x}{4}$

15) $\frac{11 + x}{5} = (-6)$

16) $(-3)(10 + 5x) = (-15)$

17) $(-3x) + 12 = 24$

18) $\frac{x + 5}{5} = -5$

19) $\frac{x + 23}{8} = 3$

20) $(-4) + \frac{x}{2} = (-14)$

21) $-5 = \frac{x + 7}{8}$

22) $\frac{9x - 3}{6} = 4$

23) $\frac{2x - 12}{8} = 6$

24) $40 = (-5)(x - 8)$

ISEE Middle-Level Subject Test Mathematics

Multi-Step Equations

✏ Find the answer for each equation.

1) $3x + 3 = 9$

2) $-x + 5 = 12$

3) $4x - 8 = 8$

4) $-(3 - x) = 5$

5) $4x - 8 = 16$

6) $12x - 15 = 9$

7) $2x - 18 = 2$

8) $4x + 8 = 16$

9) $24x + 27 = 75$

10) $-14(3 + x) = 14$

11) $-3(2 + x) = 6$

12) $12 = -(x - 7)$

13) $3(3 - x) = 30$

14) $-15 = -(3x + 6)$

15) $40(3 + x) = 40$

16) $5(x - 10) = 25$

17) $-18 = x + 8x$

18) $3x + 25 = -2x - 10$

19) $7(6 + 3x) = -63$

20) $18 - 3x = -4 - 5x$

21) $4 - 6x = 36 + 2x$

22) $15 + 15x = -5 + 5x$

23) $42 = (-6x) - 7 + 7$

24) $21 = 3x - 21 + 4x$

25) $-18 = -6x - 9 + 3x$

26) $5x - 15 = -29 + 6x$

27) $7x - 18 = 4x + 3$

28) $-7 - 4x = 5(4 - x)$

29) $x - 5 = -5(-3 - x)$

30) $13x - 68 = 15x - 102$

31) $-5x - 3 = -3(9 + 3x)$

32) $-2x - 15 = 6x + 17$

WWW.MathNotion.Com

ISEE Middle-Level Subject Test Mathematics

One-Step Inequalities

✎ Solve each inequality.

1) $7x < 14$

2) $x + 7 \geq -8$

3) $x - 1 \leq 9$

4) $-2x + 4 > -10$

5) $x + 18 \geq -6$

6) $x + 9 \geq 5$

7) $x - \frac{1}{3} \leq 5$

8) $-7x < 42$

9) $-x + 8 > -3$

10) $\frac{x}{3} + 3 > -9$

11) $-x + 8 > -4$

12) $x - 14 \leq 18$

13) $-x - 5 \leq -7$

14) $x + 26 \geq -13$

15) $x + \frac{1}{3} \geq -\frac{2}{3}$

16) $x + 6 \geq -14$

17) $x - 42 \leq -48$

18) $x - 5 \leq 4$

19) $-x + 5 > -6$

20) $x + 6 \geq -12$

21) $8x + 6 \leq 22$

22) $4x - 3 \geq 9$

23) $3x - 5 < 22$

24) $6x - 8 \leq 40$

WWW.MathNotion.Com

ISEE Middle-Level Subject Test Mathematics

Graphing Inequalities

✎ Draw a graph for each inequality.

1) $x > -1$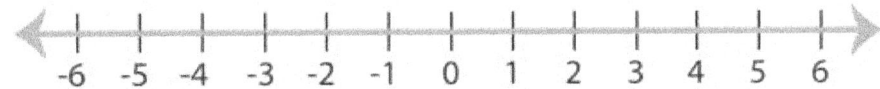

2) $x \leq 2$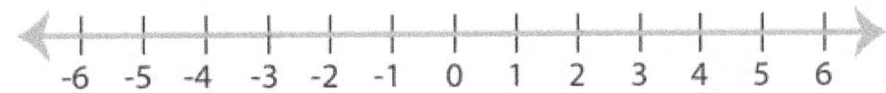

3) $x \geq 0$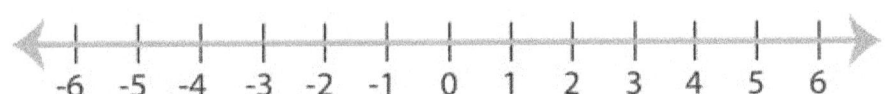

4) $x < -3$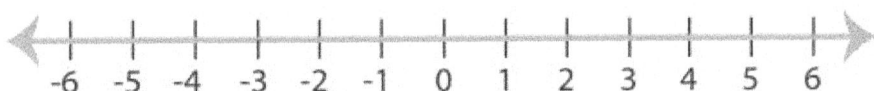

5) $x < \frac{1}{2}$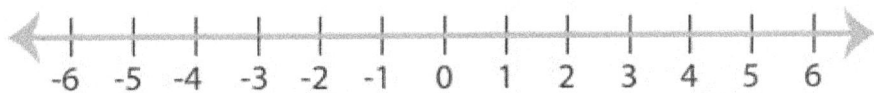

6) $x \leq -2$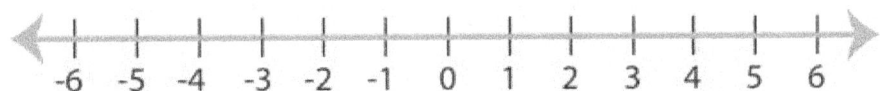

7) $x \leq 3$

8) $x \geq -\frac{7}{2}$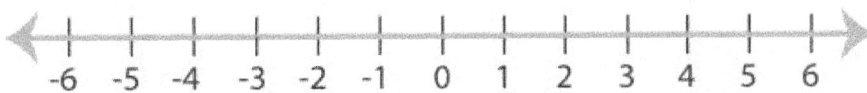

WWW.MathNotion.Com

ISEE Middle-Level Subject Test Mathematics

Two-Steps Inequality

✎ Solve each inequality

1) $2x - 3 \leq 7$

2) $3x - 4 \leq 8$

3) $\frac{-1}{4}x + \frac{x}{2} \leq \frac{1}{8}$

4) $5x + 10 \geq 30$

5) $4x - 7 \geq 9$

6) $3x - 5 \leq 16$

7) $8x - 2 \leq 14$

8) $9x + 5 \leq 23$

9) $2x + 10 > 32$

10) $\frac{x}{8} + 2 \leq 4$

11) $3x + 4 \geq 37$

12) $3x - 8 < 10$

13) $6 \geq \frac{x+7}{2}$

14) $3x + 9 < 48$

15) $\frac{4+x}{5} \geq 3$

16) $16 + 4x < 36$

17) $16 > 6x - 8$

18) $5 + \frac{x}{3} < 6$

19) $-4 + 4x > 24$

20) $5 + \frac{x}{7} < 3$

WWW.MathNotion.Com

ISEE Middle-Level Subject Test Mathematics

Multi-Step Inequalities

✎ **Calculate each inequality.**

21) $x - 3 \leq 7$

22) $8 - x \leq 8$

23) $3x - 9 \leq 9$

24) $4x - 4 \geq 8$

25) $x - 7 \geq 1$

26) $5x - 15 \leq 5$

27) $6x - 8 \leq 4$

28) $-11 + 6x \leq 12$

29) $4(x - 4) \leq 16$

30) $3x - 10 \leq 11$

31) $5x - 25 < 25$

32) $9x - 5 < 22$

33) $20 - 7x \geq -15$

34) $33 + 6x < 45$

35) $8 + 8x \geq 96$

36) $7 + 3x < 13$

37) $4x - 3 < 9$

38) $5(2 - 2x) \geq -30$

39) $-(7 + 6x) < 29$

40) $12 - 8x \geq -20$

41) $-4(x - 6) > 24$

42) $\dfrac{3x + 9}{6} \leq 10$

43) $\dfrac{4x - 10}{3} \leq 2$

44) $\dfrac{2x - 8}{3} > 2$

45) $8 + \dfrac{x}{6} < 9$

46) $\dfrac{9x}{7} - 4 < 5$

47) $\dfrac{15x + 45}{15} > 1$

48) $16 + \dfrac{x}{4} < 6$

WWW.MathNotion.Com

ISEE Middle-Level Subject Test Mathematics

Answers of Worksheets

One–Step Equations

1) 30	9) 17	17) −14	25) 22
2) 7	10) −4	18) 20	26) 9
3) 4	11) 12	19) 45	27) 60
4) 6	12) 16	20) −13	28) 35
5) 5	13) 34	21) 34	29) 54
6) 11	14) −15	22) −5	30) 24
7) 11	15) −18	23) 47	31) 30
8) 8	16) 14	24) −42	32) −21

One–Step Equation Word Problems

1) 8	3) 18	5) 5.46
2) $69.68	4) 38	6) 48.59

Two Steps Equations

1) $x = 4$	9) $x = 2$	17) $x = -4$	
2) $x = -6$	10) $x = 5$	18) $x = -30$	
3) $x = 2$	11) $x = -25$	19) $x = 1$	
4) $x = -5$	12) $x = -7$	20) $x = -20$	
5) $x = -3$	13) $x = 36$	21) $x = -47$	
6) $x = 2$	14) $x = -1$	22) $x = 3$	
7) $x = -4$	15) $x = -41$	23) $x = 30$	
8) $x = -1$	16) $x = -1$	24) $x = 0$	

Multi–Step Equations

1) 2	8) 2	15) −2	22) −2
2) −7	9) 2	16) 15	23) −7
3) 4	10) −4	17) −2	24) 6
4) 8	11) −4	18) −7	25) 3
5) 6	12) −5	19) −5	26) 14
6) 2	13) −7	20) −11	27) 7
7) 10	14) 3	21) −4	28) 27

WWW.MathNotion.Com

ISEE Middle-Level Subject Test Mathematics

29) -5 30) 17 31) -6 32) -4

One Step Inequality

1) $x < 2$
2) $x \geq -15$
3) $x \leq 10$
4) $x < 7$
5) $x \geq -24$
6) $x \geq -4$
7) $x \leq \frac{16}{3}$
8) $x > -6$
9) $x < 11$
10) $x > -36$
11) $x < 12$
12) $x \leq 32$
13) $x \geq 2$
14) $x \geq -39$
15) $x \geq -1$
16) $x \geq -20$
17) $x \leq -6$
18) $x \leq 9$
19) $x < 11$
20) $x \geq -18$
21) $x \leq 2$
22) $x \geq 3$
23) $x < 9$
24) $x \leq 8$

Graphing Single–Variable Inequalities

1)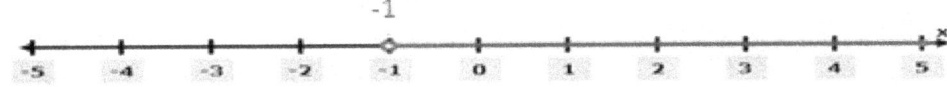

2)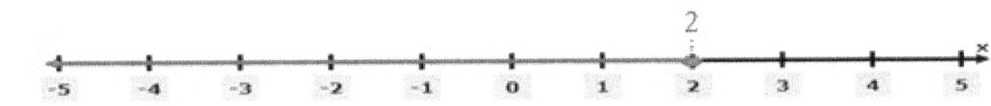

3)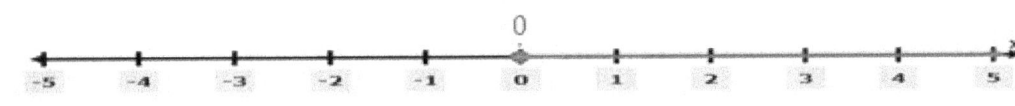

4)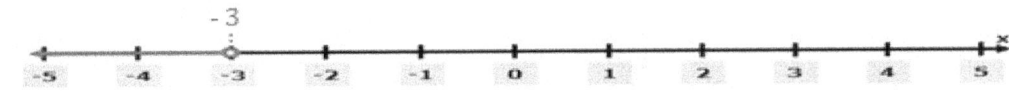

5)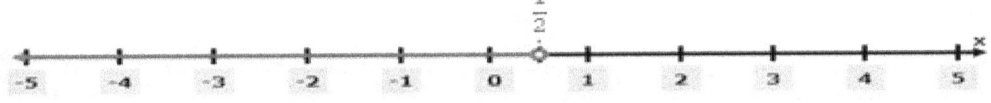

6)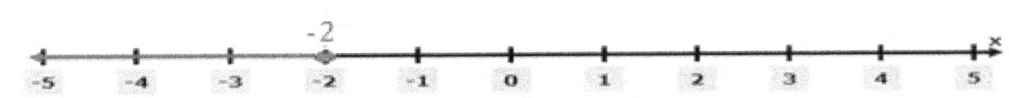

7)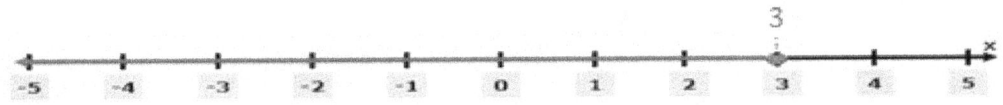

8)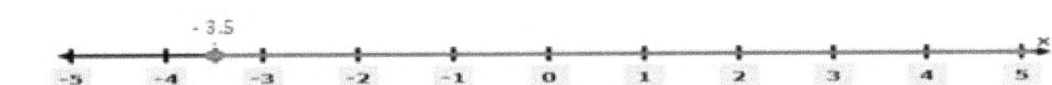

Two Steps Inequality

1) $x \leq 5$
2) $x \leq 4$
3) $x \leq 0.5$
4) $x \geq 4$
5) $x \geq 4$
6) $x \leq 7$

ISEE Middle-Level Subject Test Mathematics

7) $x \leq 2$
8) $x \leq 2$
9) $x > 11$
10) $x \leq 16$
11) $x \geq 11$

12) $x < 6$
13) $x \leq 5$
14) $x < 13$
15) $x \geq 11$
16) $x < 5$

17) $x < 4$
18) $x < 3$
19) $x > 7$
20) $x < -14$

Multi-Step Inequalities

1) $x \leq 10$
2) $x \geq 0$
3) $x \leq 6$
4) $x \geq 3$
5) $x \geq 8$
6) $x \leq 4$
7) $x \leq 2$

8) $x \leq \frac{23}{6}$
9) $x \leq 8$
10) $x \leq 7$
11) $x < 10$
12) $x < 3$
13) $x \leq 5$
14) $x < 2$

15) $x \geq 11$
16) $x < 2$
17) $x < 3$
18) $x \leq 4$
19) $x > -6$
20) $x \leq 4$
21) $x < 0$

22) $x \leq 17$
23) $x \leq 4$
24) $x > 7$
25) $x < 6$
26) $x < 7$
27) $x > -2$
28) $x < -40$

ISEE Middle-Level Subject Test Mathematics

Chapter 10 :

Geometry and Solid Figures

Topics that you will practice in this chapter:

- ✓ Angles
- ✓ Pythagorean Relationship
- ✓ Triangles
- ✓ Polygons
- ✓ Trapezoids
- ✓ Circles
- ✓ Cubes
- ✓ Rectangular Prism
- ✓ Cylinder

Mathematics is, as it were, a sensuous logic, and relates to philosophy as do the arts, music, and plastic art to poetry. — *K. Shegel*

ISEE Middle-Level Subject Test Mathematics

Angles

✎ **What is the value of x in the following figures?**

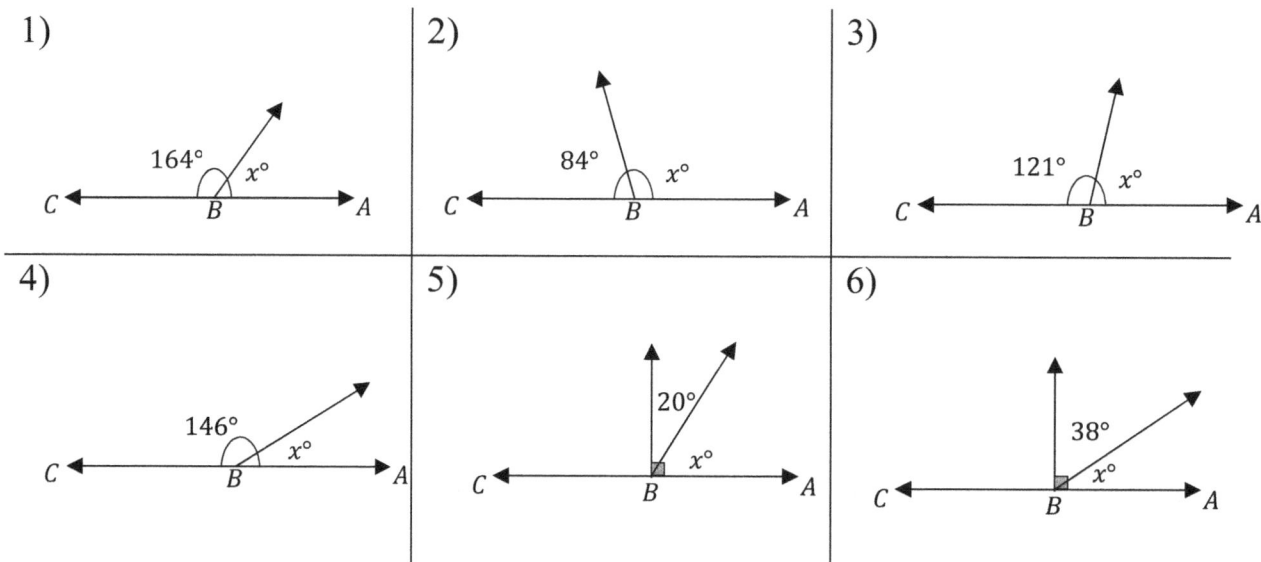

✎ **Calculate.**

7) Two supplement angles have equal measures. What is the measure of each angle? _____

8) The measure of an angle is seven fifth the measure of its supplement. What is the measure of the angle? _____

9) Two angles are complementary and the measure of one angle is 24 less than the other. What is the measure of the smaller angle? _____

10) Two angles are complementary. The measure of one angle is one fifth the measure of the other. What is the measure of the bigger angle? _____

11) Two supplementary angles are given. The measure of one angle is 40° less than the measure of the other. What does the smaller angle measure? _____

WWW.MathNotion.Com

ISEE Middle-Level Subject Test Mathematics

Pythagorean Relationship

✎ Do the following lengths form a right triangle?

1)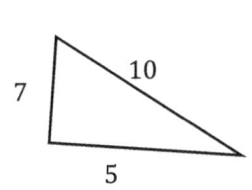

2)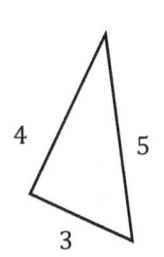

3)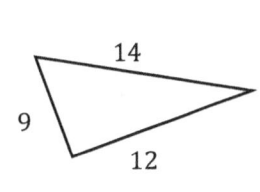

4)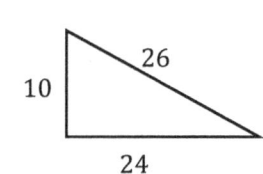

5)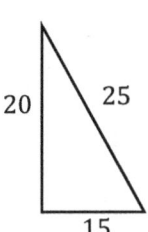

6)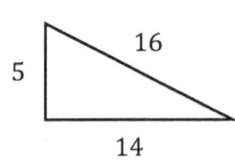

7)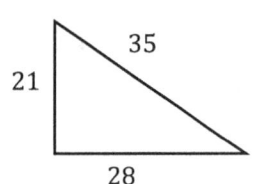

8)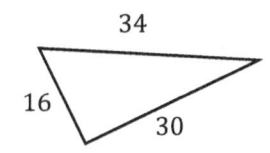

✎ Find the missing side?

9)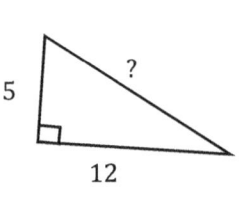

10)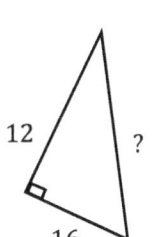

11)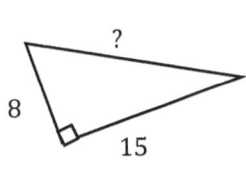

12)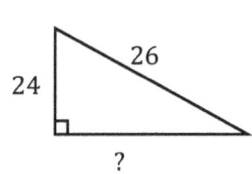

13)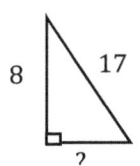

14)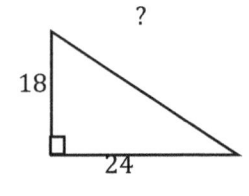

15)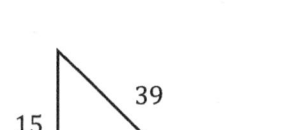

16)

WWW.MathNotion.Com   107

ISEE Middle-Level Subject Test Mathematics

Triangles

✎ Find the measure of the unknown angle in each triangle.

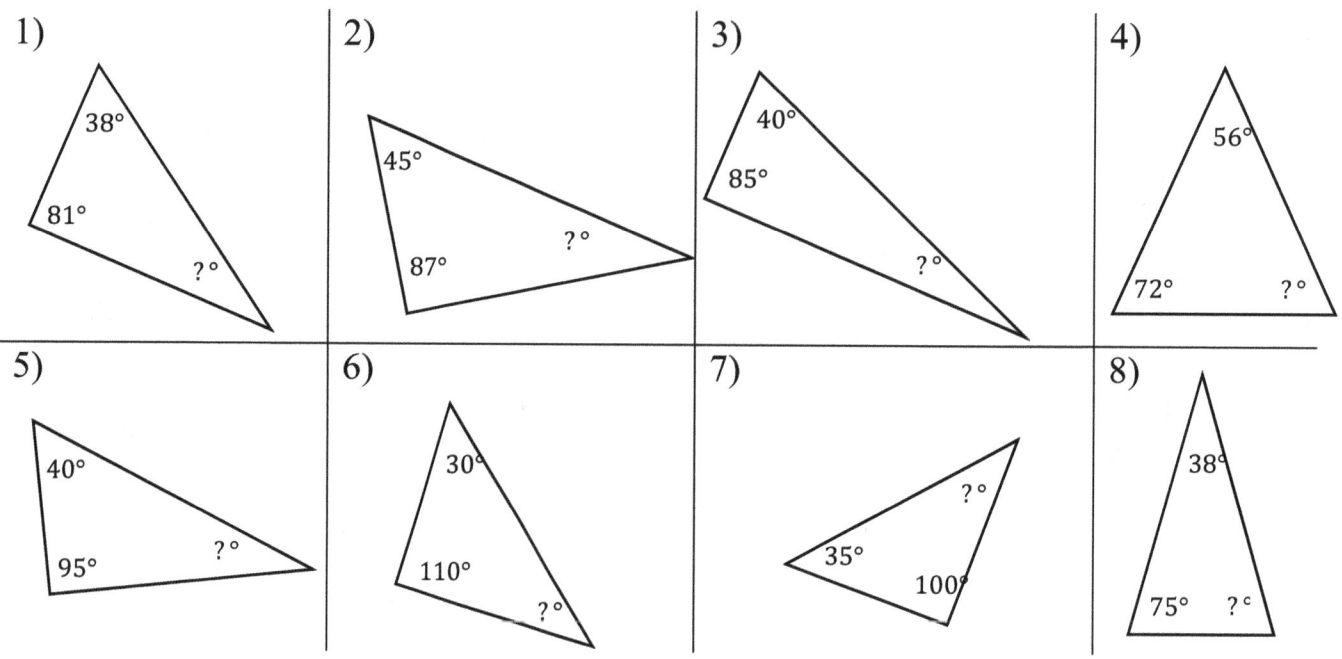

✎ Find area of each triangle.

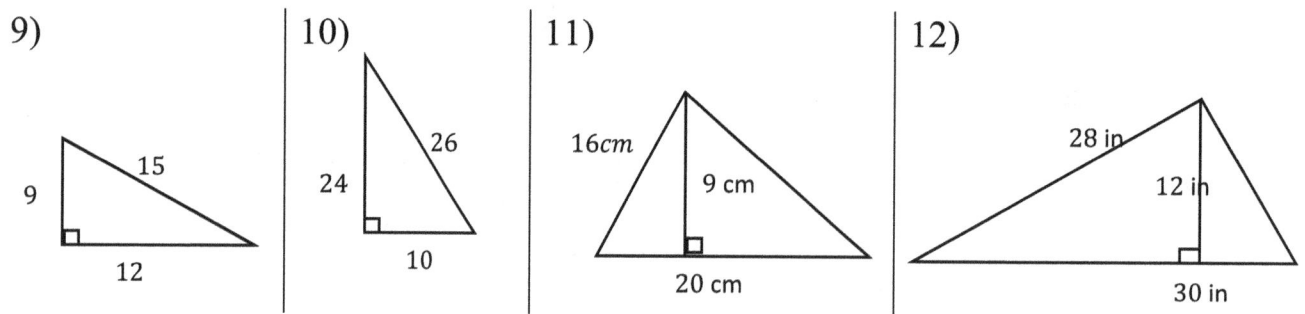

WWW.MathNotion.Com

ISEE Middle-Level Subject Test Mathematics

Polygons

✍ **Find the perimeter of each shape.**

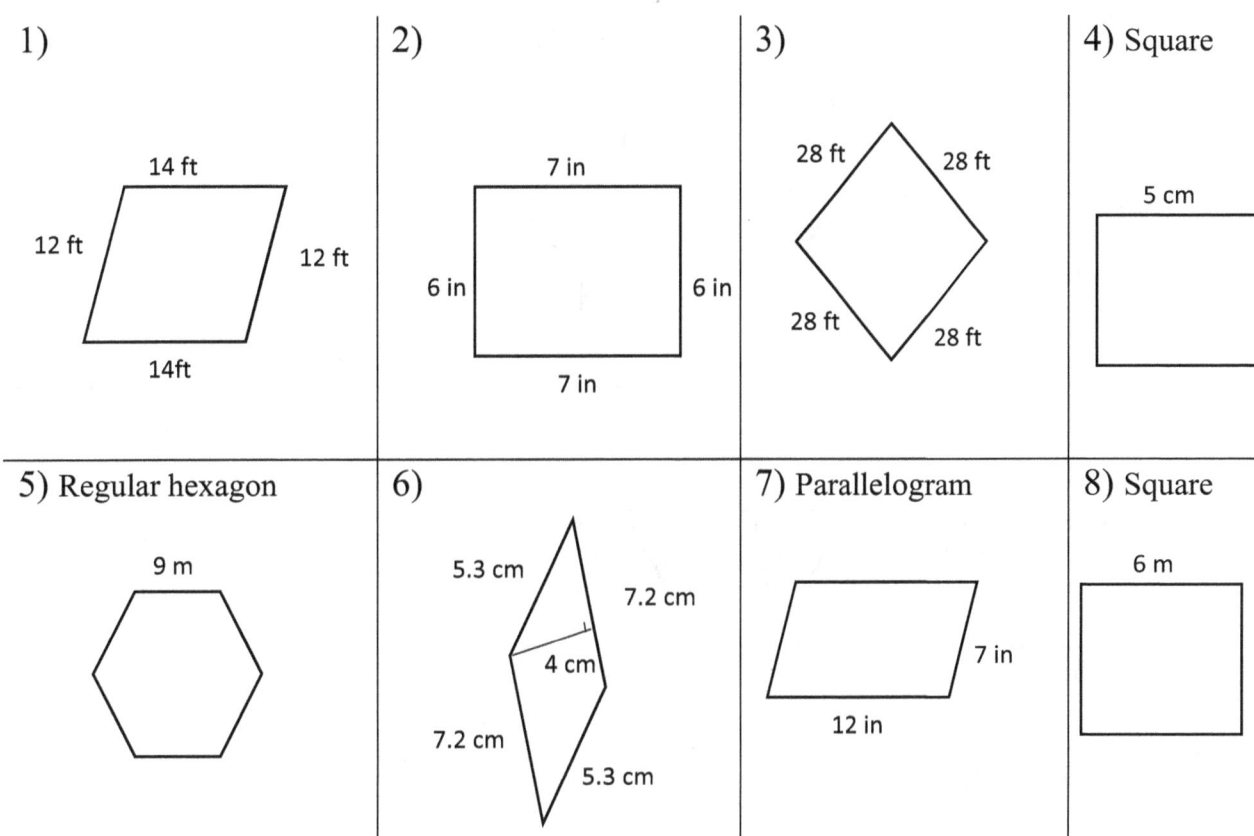

✍ **Find the area of each shape.**

9) Parallelogram

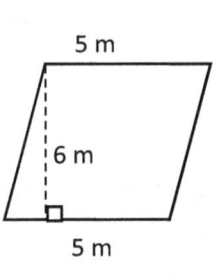

10) Rectangle

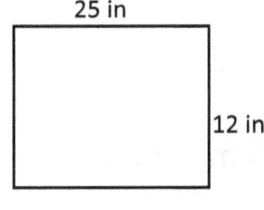

11) Rectangle

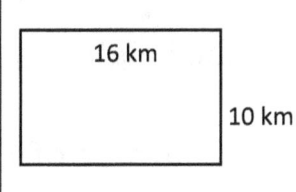

12) Square

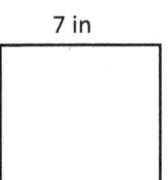

ISEE Middle-Level Subject Test Mathematics

Trapezoids

✎ Find the area of each trapezoid.

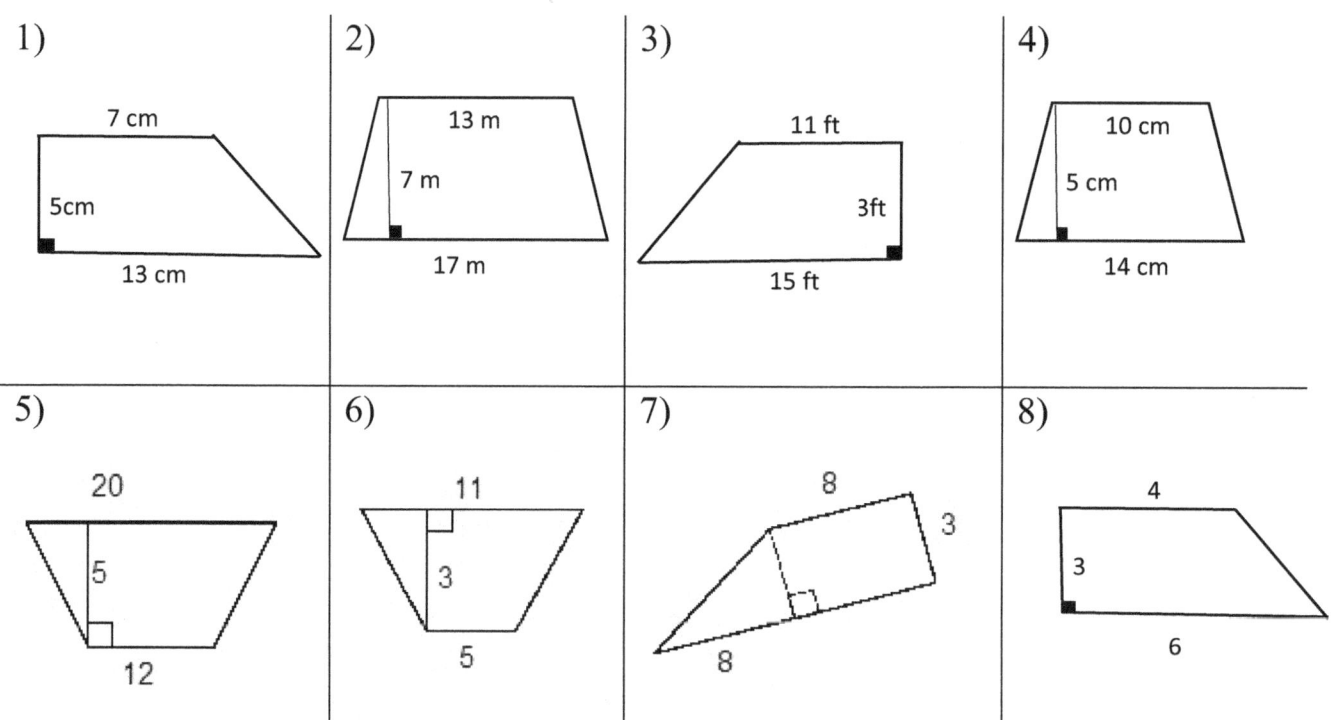

✎ Calculate.

1) A trapezoid has an area of 45 cm² and its height is 5 cm and one base is 5 cm. What is the other base length? _____

2) If a trapezoid has an area of 99 ft² and the lengths of the bases are 8 ft and 10 ft, find the height? _____

3) If a trapezoid has an area of 126 m² and its height is 14 m and one base is 6 m, find the other base length? _____

4) The area of a trapezoid is 440 ft² and its height is 22 ft. If one base of the trapezoid is 15 ft, what is the other base length? _____

ISEE Middle-Level Subject Test Mathematics

Circles

✏️ **Find the area of each circle.** ($\pi = 3.14$)

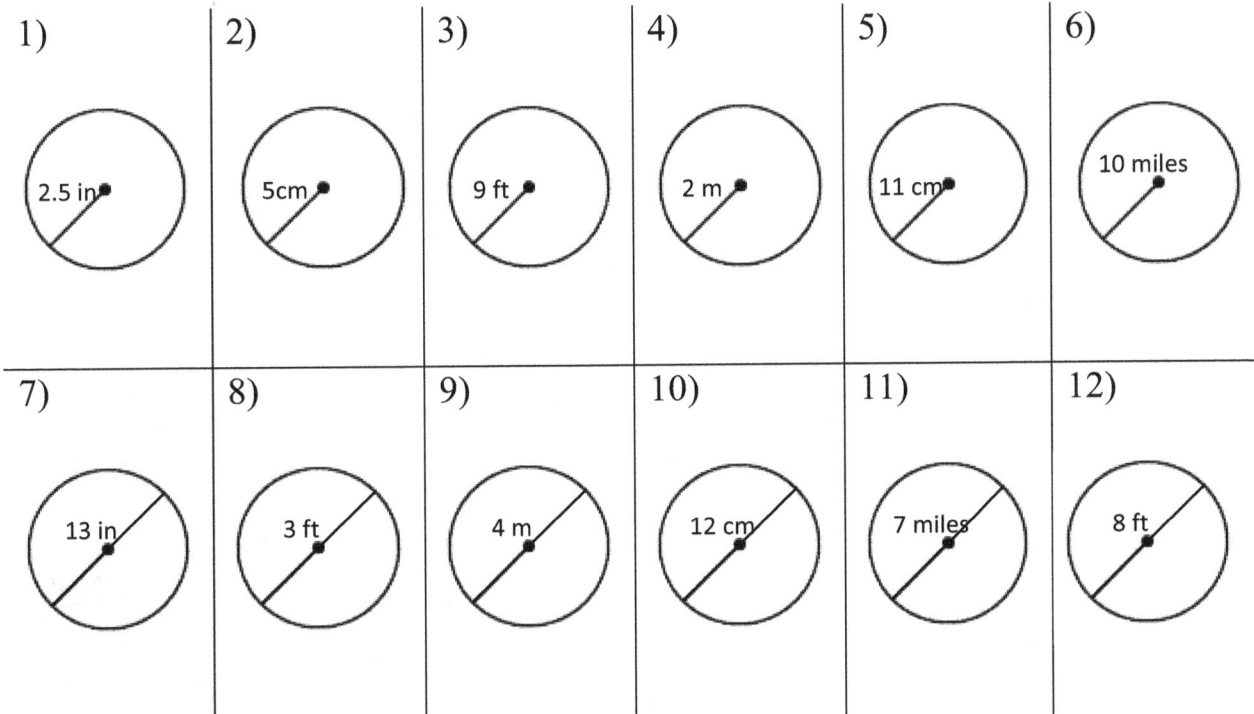

✏️ **Complete the table below.** ($\pi = 3.14$)

Circle No.	Radius	Diameter	Circumference	Area
1	1 in	2 in	6.28 in	3.14 in^2
2		10 m		
3				28.26 ft^2
4			47.1 mi	
5		11 km		
6	7 cm			
7		12 ft		
8				314 m^2
9			56.52 in	
10	4.5 ft			

ISEE Middle-Level Subject Test Mathematics

Cubes

✏️ **Find the volume of each cube.**

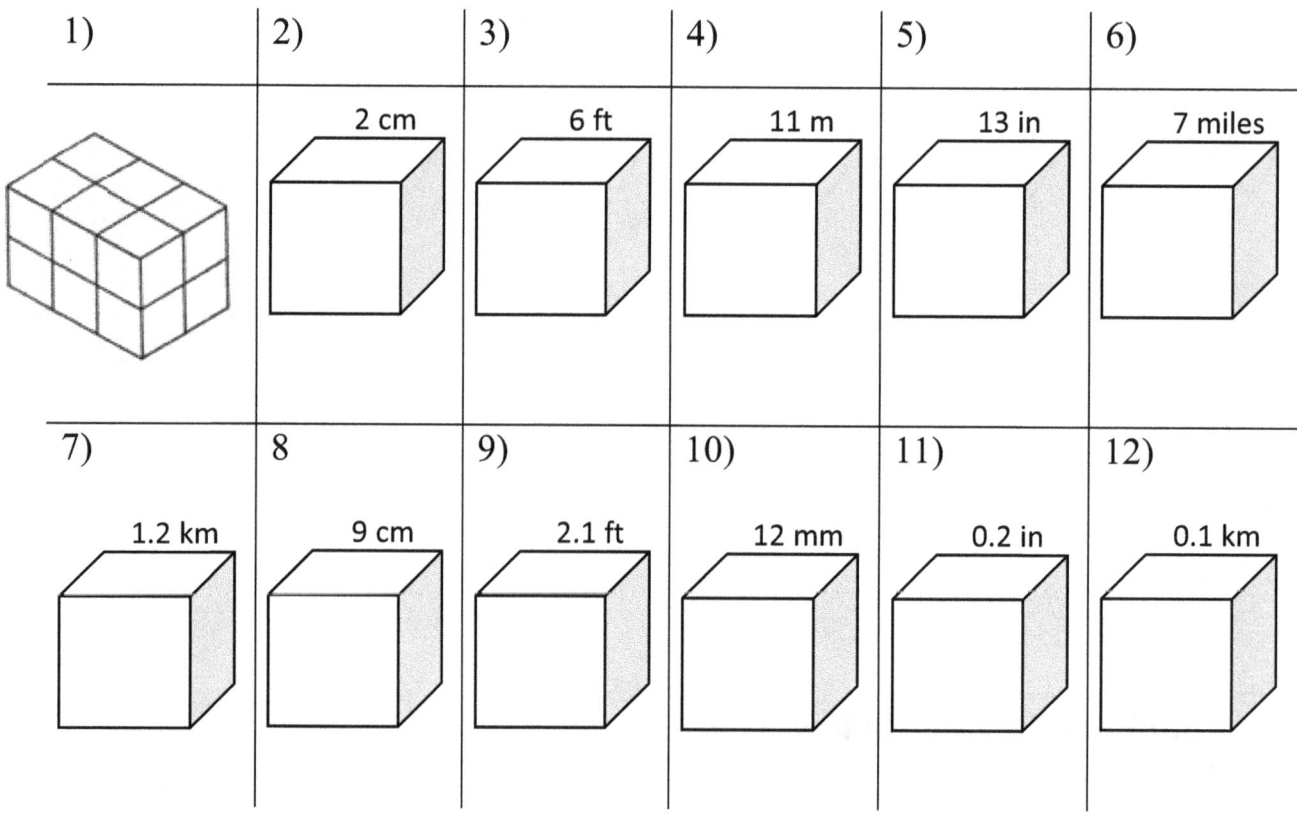

✏️ **Find the surface area of each cube.**

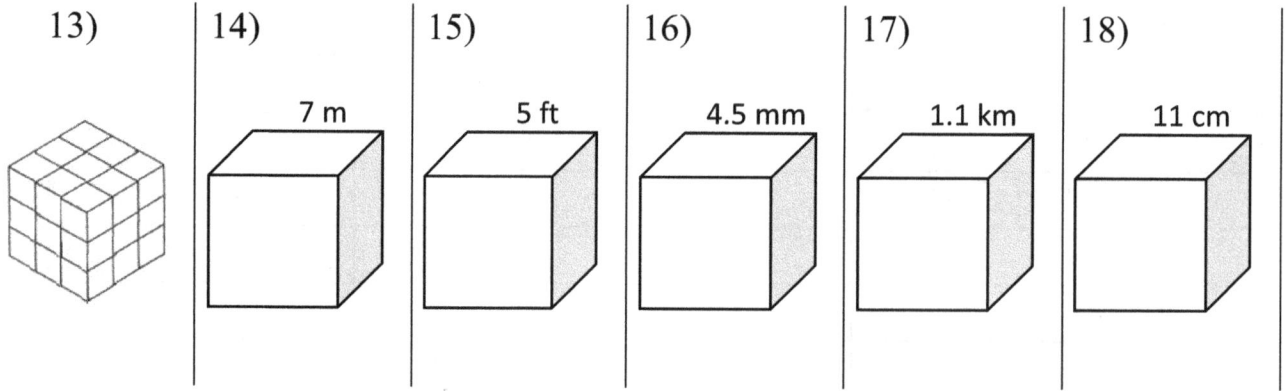

WWW.MathNotion.Com

ISEE Middle-Level Subject Test Mathematics

Rectangular Prism

✍ Find the volume of each Rectangular Prism.

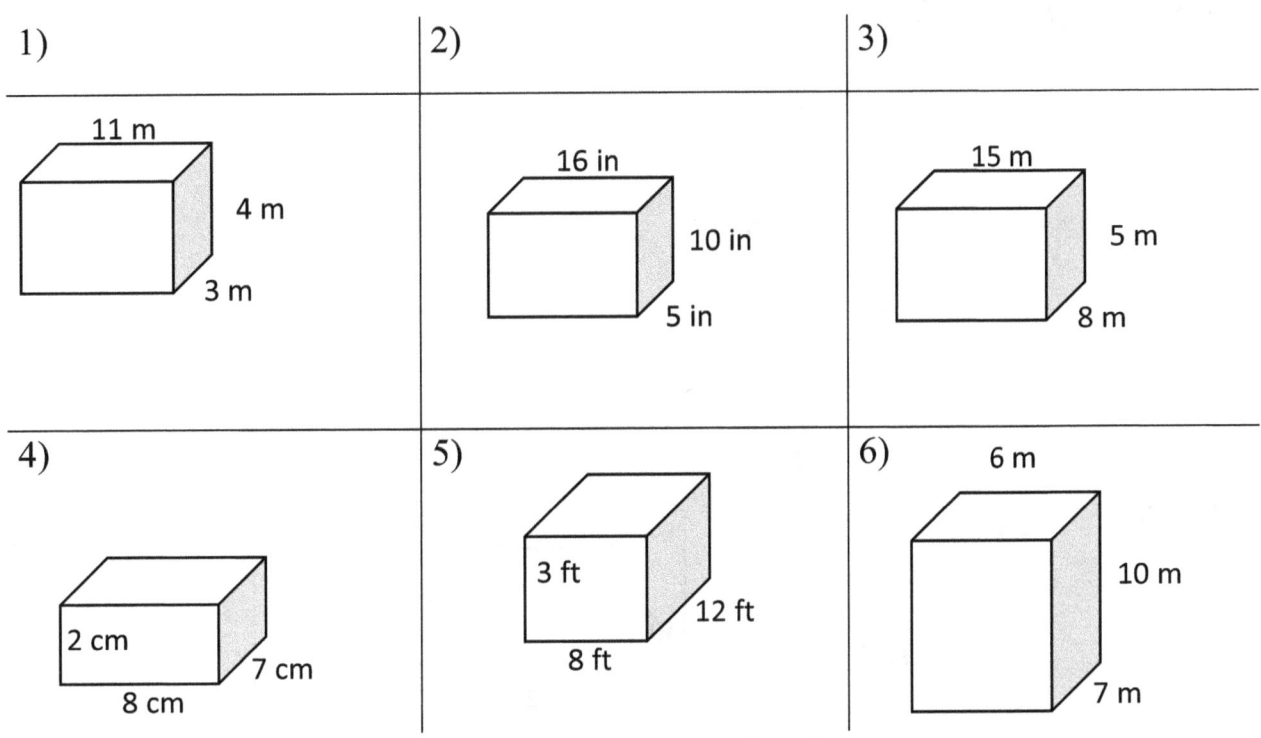

✍ Find the surface area of each Rectangular Prism.

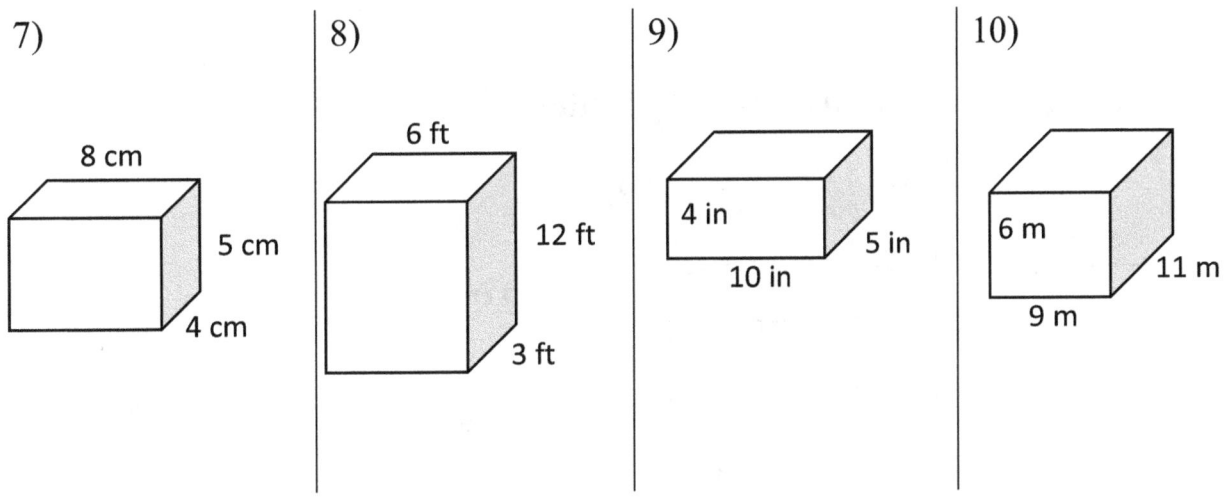

WWW.MathNotion.Com

ISEE Middle-Level Subject Test Mathematics

Cylinder

✏️ **Find the volume of each Cylinder. Round your answer to the nearest tenth.** ($\pi = 3.14$)

1)
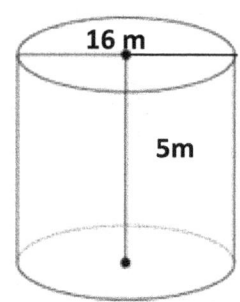

2)
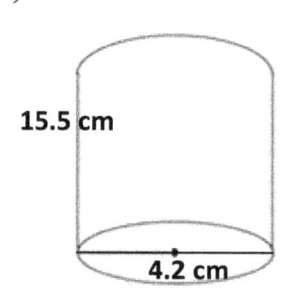

3)
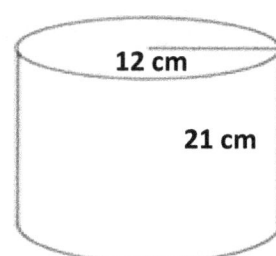

4)
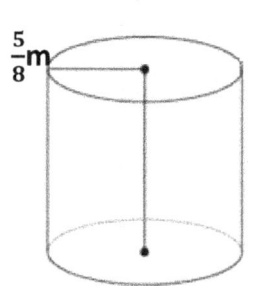

5)
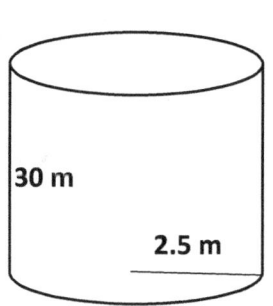

6)
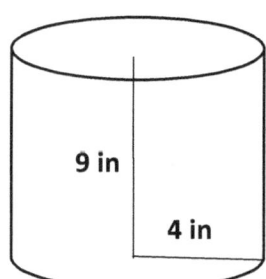

✏️ **Find the surface area of each Cylinder.** ($\pi = 3.14$)

7)
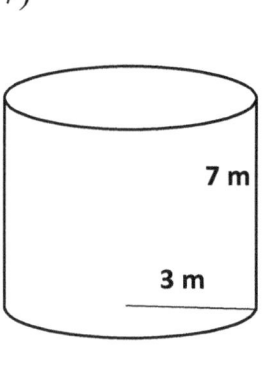

8)
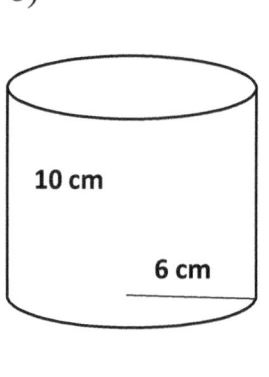

9)
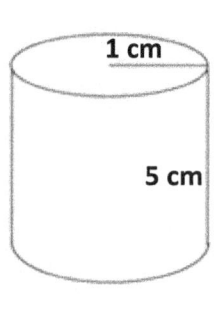

10)
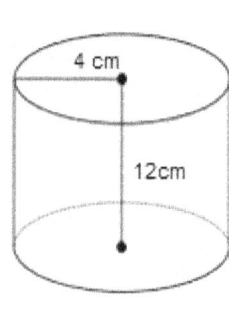

WWW.MathNotion.Com

ISEE Middle-Level Subject Test Mathematics

Answers of Worksheets

Angles
1) 16°
2) 96°
3) 59°
4) 34°
5) 70°
6) 52°
7) 90°
8) 75°
9) 33°
10) 75°
11) 70°

Pythagorean Relationship
1) No
2) Yes
3) No
4) Yes
5) Yes
6) No
7) Yes
8) Yes
9) 13
10) 20
11) 17
12) 10
13) 15
14) 30
15) 36
16) 12

Triangles
1) 60°
2) 48°
3) 55°
4) 52°
5) 45°
6) 40°
7) 45°
8) 67°
9) 54 square unites
10) 120 square unites
11) 90 square unites
12) 180 square unites

Polygons
1) 52 ft
2) 26 in
3) 112 ft
4) 20 cm
5) 54 m
6) 25 cm
7) 38 in
8) 24 m
9) 30 m^2
10) 300 in^2
11) 160 km^2
12) 49 in^2

Trapezoids
1) 50 cm^2
2) 105 m^2
3) 39 ft^2
4) 60 cm^2
5) 80
6) 24
7) 36
8) 15

Calculate
1) 13 cm
2) 11 ft
3) 12 m
4) 25 ft

Circles
1) 19.63 in^2
2) 78.5 cm^2
3) 254.34 ft^2
4) 12.56 m^2
5) 379.94 cm^2
6) 314 $miles^2$
7) 132.67 in^2
8) 7.07 ft^2
9) 12.56 m^2
10) 113.04 cm^2
11) 38.47 $miles^2$
12) 50.24 ft^2

ISEE Middle-Level Subject Test Mathematics

Circle No.	Radius	Diameter	Circumference	Area
1	1 in	2 in	6.28 in	3.14 in^2
2	5 m	10 m	31.4 m	78.5 m^2
3	3 ft	6 ft	18.84 ft	28.26 ft^2
4	7.5 miles	15 mi	47.1 mi	176.63 mi^2
5	5.5 km	11 km	34.54 km	94.99 km^2
6	7 cm	14 cm	43.96 cm	153.86 cm^2
7	6 ft	12 ft	37.68 feet	113.04 ft^2
8	10 m	20 m	62.8 m	314 m^2
9	9 in	18 in	56.52 in	254.34 in^2
10	4.5 ft	9 ft	28.26 ft	63.585 ft^2

Cubes

1) 12
2) 8 cm^3
3) 216 ft^3
4) 1,331 m^3
5) 2,197 in^3
6) 343 $miles^3$
7) 1.728 km^3
8) 729 cm^3
9) 9.261 ft^3
10) 1,728 mm^3
11) 0.008 in^3
12) 0.001 km^3
13) 27
14) 294 m^2
15) 150 ft^2
16) 121.5 mm^2
17) 7.26 km^2
18) 726 cm^2

Rectangular Prism

1) 132 m^3
2) 800 in^3
3) 600 m^3
4) 112 cm^3
5) 288 ft^3
6) 420 m^3
7) 184 cm^2
8) 252 ft^2
9) 220 in^2
10) 438 m^2

Cylinder

1) 1,004.8 m^3
2) 214.6 cm^3
3) 9,495.4 cm^3
4) 1.1 m^3
5) 588.8 m^3
6) 452.2 in^3
7) 188.4 m^2
8) 602.9 cm^2
9) 37.7 cm^2
10) 401.9 m^2

WWW.MathNotion.Com

ISEE Middle-Level Subject Test Mathematics

Chapter 11 :
Statistics and Probability

Topics that you will practice in this chapter:

- ✓ Mean and Median
- ✓ Mode and Range
- ✓ Times Series
- ✓ Stem–and–Leaf Plot
- ✓ Pie Graph
- ✓ Probability Problems

Mathematics is no more computation than typing is literature.
– John Allen Paulos

ISEE Middle-Level Subject Test Mathematics

Mean and Median

✎ **Find Mean and Median of the Given Data.**

1) 8, 7, 14, 4, 8

2) 14, 8, 25, 19, 16, 33, 11

3) 23, 18, 15, 12, 17

4) 34, 14, 10, 15, 6, 11

5) 10, 19, 6, 8, 32, 20, 17

6) 17, 26, 39, 69, 20, 6

7) 40, 38, 18, 11, 9, 2, 7, 32, 41

8) 24, 21, 31, 12, 33, 32, 22

9) 16, 14, 20, 41, 15, 20, 38, 4

10) 20, 20, 30, 18, 6, 28, 12, 46

11) 12, 7, 10, 11, 16, 22

12) 10, 29, 27, 12, 2, 15, 10, 3

✎ **Calculate.**

13) In a javelin throw competition, five athletics score 56, 34, 62, 23 and 19 meters. What are their Mean and Median? _____

14) Eva went to shop and bought 8 apples, 14 peaches, 6 bananas, 4 pineapples and 12 melons. What are the Mean and Median of her purchase? _____

15) Bob has 17 black pen, 19 red pen, 14 green pens, 20 blue pens and 5 boxes of yellow pens. If the Mean and Median are 19 respectively, what is the number of yellow pens in each box? _____

ISEE Middle-Level Subject Test Mathematics

Mode and Range

✏️ **Find Mode and Rage of the Given Data.**

1) 4, 3, 7, 3, 3, 4
   Mode: _____   Range: _____

2) 18, 18, 24, 26, 18, 8, 14, 22
   Mode: _____   Range: _____

3) 8, 8, 8, 16, 19, 22, 20, 9, 13
   Mode: _____   Range: _____

4) 24, 24, 14, 28, 20, 18, 20, 24
   Mode: _____   Range: _____

5) 6, 21, 27, 24, 27, 27
   Mode: _____   Range: _____

6) 21, 8, 8, 7, 8, 12, 10, 22, 18, 13
   Mode: _____   Range: _____

7) 7, 4, 4, 6, 13, 13, 13, 0, 2, 2
   Mode: _____   Range: _____

8) 5, 8, 5, 14, 12, 14, 3, 5, 18
   Mode: _____   Range: _____

9) 7, 7, 7, 12, 7, 3, 8, 16, 3, 17
   Mode: _____   Range: _____

10) 15, 15, 19, 16, 4, 16, 10, 15
    Mode: _____   Range: _____

11) 6, 6, 5, 6, 42, 13, 19, 2
    Mode: _____   Range: _____

12) 8, 8, 9, 8, 9, 4, 34, 22
    Mode: _____   Range: _____

✏️ **Calculate.**

13) A stationery sold 12 pencils, 56 red pens, 24 blue pens, 20 notebooks, 12 erasers, 21 rulers and 11 color pencils. What are the Mode and Range for the stationery sells?

   Mode: _____   Range: _____

14) In an English test, eight students score 10, 15, 15, 18 18, 16, 15 and 15. What are their Mode and Range? _____

15) What is the range of the first 6 even numbers greater than 8?
   _____

WWW.MathNotion.Com

ISEE Middle-Level Subject Test Mathematics

Times Series

✏️ **Use the following Graph to complete the table.**

Day	Distance (km)
1	
2	

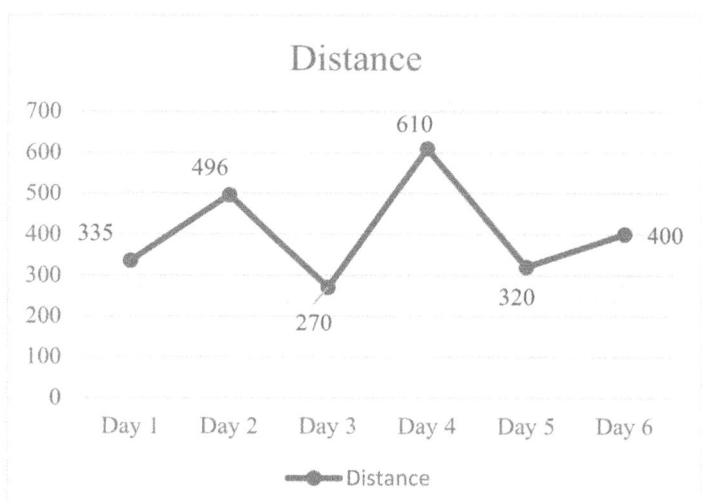

The following table shows the number of births in the US from 2007 to 2012 (in millions).

Year	Number of births (in millions)
2007	4.15
2008	3.70
2009	3.45
2010	3.20
2011	1.75
2012	2.98

Draw a Time Series for the table.

ISEE Middle-Level Subject Test Mathematics

Stem–and–Leaf Plot

✍ **Make stem ad leaf plots for the given data.**

1) 24, 26, 29, 20, 53, 27, 51, 55, 36, 21, 37, 30   Stem | Leaf plot

2) 11, 59, 66, 14, 18, 19, 59, 65, 69, 61, 68, 65   Stem | Leaf plot

3) 121, 55, 66, 54, 112, 128, 63, 125, 59, 123, 68, 119   Stem | Leaf plot

4) 51, 32, 100, 56, 84, 36, 107, 56, 85, 39, 56, 106, 89 Stem | Leaf plot

5) 33, 89, 19, 87, 81, 16, 11, 30, 86, 35, 17, 35, 13   Stem | Leaf plot

6) 60, 92, 22, 25, 67, 93, 95, 62, 21, 64, 98, 29   Stem | Leaf plot

WWW.MathNotion.Com

ISEE Middle-Level Subject Test Mathematics

Quartile of a Data Set

✏️ **Find First, Second and Third Quartile of the Given Data.**

1) 45, 8, 25, 43, 24, 36, 35, 62, 19

2) 23, 63, 25, 19, 80, 32

3) 86, 33, 85, 60, 72, 42, 51, 46

4) 24, 44, 48, 25, 25, 36, 25, 36, 71, 49

5) 23, 15, 45, 9, 35, 8, 25, 15

6) 66, 86, 40, 32, 82, 25, 52, 44, 61

Box and Whisker Plots

✏️ **Make box and whisker plots for the given data.**

1) 86, 65, 92, 67, 72, 87, 87, 83, 95, 66, 76, 82

2) 8, 22, 17, 15, 13, 5, 8, 12, 6, 11, 6, 15, 4, 28

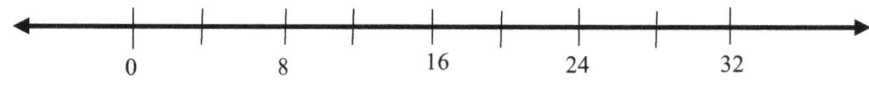

3) 25, 21, 34, 19, 23, 24, 13, 17, 15, 16, 22

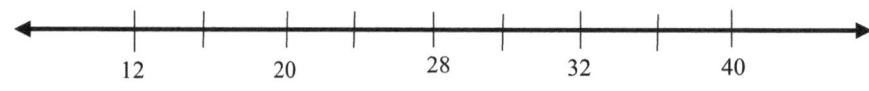

ISEE Middle-Level Subject Test Mathematics

Pie Graph

The circle graph below shows all Robert's expenses for last month. Robert spent $140 on his hobbies last month.

Answer following questions based on the Pie graph.

1) How much was Robert's total expenses last month? _____

2) How much did Robert spend on his car last month? _____

3) How much did Robert spend for shopping last month? _____

4) How much did Robert spend on his rent last month? _____

5) What fraction is Robert's expenses for his rent and car out of his total expenses last month? _____

WWW.MathNotion.Com

ISEE Middle-Level Subject Test Mathematics

Probability Problems

✎ **Calculate.**

1) A number is chosen at random from 1 to 10. Find the probability of selecting number 6 or smaller numbers. _____

2) Bag A contains 18 red marbles and 6 green marbles. Bag B contains 16 black marbles and 8 orange marbles. What is the probability of selecting a green marble at random from bag A? What is the probability of selecting a black marble at random from Bag B? _____

3) A number is chosen at random from 1 to 20. What is the probability of selecting multiples of 4? _____

4) A card is chosen from a well-shuffled deck of 52 cards. What is the probability that the card will be a queen? _____

5) A number is chosen at random from 1 to 15. What is the probability of selecting a multiple of 3 or 5? _____

A spinner numbered 1–8, is spun once. What is the probability of spinning …?

6) an Odd number? _____   7) a multiple of 2? _____

8) a multiple of 5? _____   9) number 10? _____

WWW.MathNotion.Com

ISEE Middle-Level Subject Test Mathematics

Answers of Worksheets

Mean and Median

1) Mean: 8.2, Median: 8
2) Mean: 18, Median: 16
3) Mean: 17, Median: 17
4) Mean: 15, Median: 12.5
5) Mean: 16, Median: 17
6) Mean: 29.5, Median: 23
7) Mean: 22, Median: 18
8) Mean: 25, Median: 24
9) Mean: 21, Median: 18
10) Mean: 22.5, Median: 20
11) Mean: 13, Median: 11.5
12) Mean: 13.5, Median: 11
13) Mean: 38.8, Median: 34
14) Mean: 8.8, Median: 8
15) 5

Mode and Range

1) Mode: 3, Range: 4
2) Mode: 18, Range: 18
3) Mode: 8, Range: 14
4) Mode: 24, Range: 14
5) Mode: 27, Range: 21
6) Mode: 8, Range: 15
7) Mode: 13, Range: 13
8) Mode: 5, Range: 15
9) Mode: 7, Range: 14
10) Mode: 15, Range: 15
11) Mode: 6, Range: 40
12) Mode: 8, Range: 30
13) Mode: 12, Range: 45
14) Mode: 15, Range: 8
15) 10

Time series

Day	Distance (km)
1	335
2	496
3	270
4	610
5	320
6	400

Stem–And–Leaf Plot

1)

Stem	leaf
2	0 1 4 6 7 9
3	0 6 7
5	1 3 5

2)

Stem	leaf
1	1 4 8 9
5	9 9
6	1 5 5 6 8 9

3)

Stem	leaf
5	4 5 9
6	3 6 8
11	2 9
12	1 3 5 8

WWW.MathNotion.Com

ISEE Middle-Level Subject Test Mathematics

4)		5)		6)	
Stem	leaf	Stem	leaf	Stem	leaf
3	2 6 9	1	1 3 6 7 9	2	2 1 5 9
5	1 6 6 6	3	0 3 5 5	6	0 2 4 7
8	4 5 9	8	1 6 7 9	9	2 3 5 8
10	0 6 7				

Quartile of the Given Data

1) First quartile: 21.5, second quartile: 35, third quartile: 44
2) First quartile: 22, second quartile: 28.5, third quartile: 67.25
3) First quartile: 43, second quartile: 55.5, third quartile: 81.75
4) First quartile: 25, second quartile: 36, third quartile: 48.25
5) First quartile: 10.5, second quartile: 19, third quartile: 32.5
6) First quartile: 36, second quartile: 52, third quartile: 74

Box and Whisker Plots

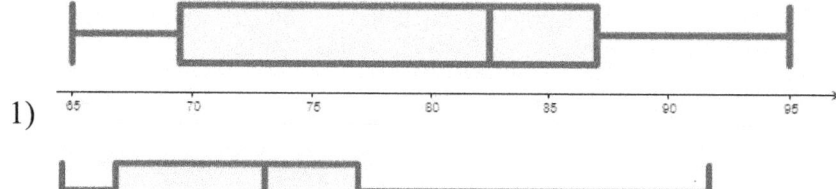
1)

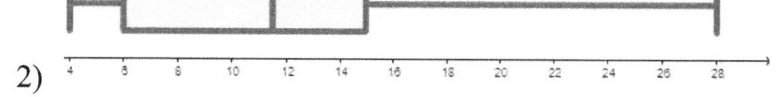
2)

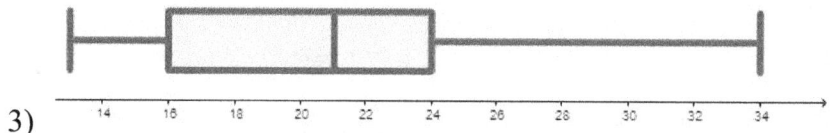
3)

Pie Graph

1) $5,600
2) $784
3) $1,288
4) $2,016
5) $\frac{1}{2}$

Probability Problems

1) $\frac{3}{5}$
2) $\frac{1}{4}, \frac{2}{3}$
3) $\frac{1}{4}$
4) $\frac{1}{13}$
5) $\frac{7}{15}$
6) $\frac{1}{2}$
7) $\frac{1}{2}$
8) $\frac{1}{8}$
9) 0

WWW.MathNotion.Com

ISEE Middle-Level Subject Test Mathematics

Chapter 12 : ISEE Middle Level Practice Tests

The Independent School Entrance Exam (ISEE) is an admission test developed by the Educational Records Bureau for its member schools as part of their admission process.

ISEE Middle Level tests use a multiple-choice format and contain two Mathematics sections:

Quantitative Reasoning:

There are 37 questions in the Quantitative Reasoning section and students have 35 minutes to answer the questions. This section contains word problems and quantitative comparisons. The word problems require either no calculation or simple calculation. The quantitative comparison items present two quantities, (A) and (B), and the student needs to select one of the following four answer choices:

(A) The quantity in Column A is greater.

(B) The quantity in Column B is greater.

(C) The two quantities are equal.

(D) The relationship cannot be determined from the information given.

Mathematics Achievement:

There are 47 questions in the Mathematics Achievement section and students have 40 minutes to answer the questions. Mathematics Achievement measures students' knowledge of Mathematics requiring one or more steps in calculating the answer.

In this section, there are two complete ISEE Middle Level Quantitative Reasoning and Mathematics Achievement Tests. Let your student take these tests to see what score they'll be able to receive on a real ISEE test.

Time to Test

Time to refine your skill with a practice examination.

Take a practice ISEE Middle Level Math Test to simulate the test day experience. After you've finished, score your test using the answer key.

Before You Start

- You'll need a pencil and scratch papers to take the test.
- For each question, there are four possible answers. Choose which one is best.
- It's okay to guess. You won't lose any points if you're wrong.
- Use the answer sheet provided to record your answers.
- After you've finished the test, review the answer key to see where you went wrong.
- **Calculators are NOT allowed for the ISEE Middle Level Test.**

Good Luck!

ISEE Middle Level Practice Test Answer Sheets

Remove (or photocopy) these answer sheets and use them to complete the practice tests.

ISEE Middle Level Practice Test

Quantitative Reasoning

#		#	
1	Ⓐ Ⓑ Ⓒ Ⓓ	21	Ⓐ Ⓑ Ⓒ Ⓓ
2	Ⓐ Ⓑ Ⓒ Ⓓ	22	Ⓐ Ⓑ Ⓒ Ⓓ
3	Ⓐ Ⓑ Ⓒ Ⓓ	23	Ⓐ Ⓑ Ⓒ Ⓓ
4	Ⓐ Ⓑ Ⓒ Ⓓ	24	Ⓐ Ⓑ Ⓒ Ⓓ
5	Ⓐ Ⓑ Ⓒ Ⓓ	25	Ⓐ Ⓑ Ⓒ Ⓓ
6	Ⓐ Ⓑ Ⓒ Ⓓ	26	Ⓐ Ⓑ Ⓒ Ⓓ
7	Ⓐ Ⓑ Ⓒ Ⓓ	27	Ⓐ Ⓑ Ⓒ Ⓓ
8	Ⓐ Ⓑ Ⓒ Ⓓ	28	Ⓐ Ⓑ Ⓒ Ⓓ
9	Ⓐ Ⓑ Ⓒ Ⓓ	29	Ⓐ Ⓑ Ⓒ Ⓓ
10	Ⓐ Ⓑ Ⓒ Ⓓ	30	Ⓐ Ⓑ Ⓒ Ⓓ
11	Ⓐ Ⓑ Ⓒ Ⓓ	31	Ⓐ Ⓑ Ⓒ Ⓓ
12	Ⓐ Ⓑ Ⓒ Ⓓ	32	Ⓐ Ⓑ Ⓒ Ⓓ
13	Ⓐ Ⓑ Ⓒ Ⓓ	33	Ⓐ Ⓑ Ⓒ Ⓓ
14	Ⓐ Ⓑ Ⓒ Ⓓ	34	Ⓐ Ⓑ Ⓒ Ⓓ
15	Ⓐ Ⓑ Ⓒ Ⓓ	35	Ⓐ Ⓑ Ⓒ Ⓓ
16	Ⓐ Ⓑ Ⓒ Ⓓ	36	Ⓐ Ⓑ Ⓒ Ⓓ
17	Ⓐ Ⓑ Ⓒ Ⓓ	37	Ⓐ Ⓑ Ⓒ Ⓓ
18	Ⓐ Ⓑ Ⓒ Ⓓ	38	Ⓐ Ⓑ Ⓒ Ⓓ
19	Ⓐ Ⓑ Ⓒ Ⓓ	39	Ⓐ Ⓑ Ⓒ Ⓓ
20	Ⓐ Ⓑ Ⓒ Ⓓ	40	Ⓐ Ⓑ Ⓒ Ⓓ

Mathematics Achievement

#		#		#	
1	Ⓐ Ⓑ Ⓒ Ⓓ	21	Ⓐ Ⓑ Ⓒ Ⓓ	41	Ⓐ Ⓑ Ⓒ Ⓓ
2	Ⓐ Ⓑ Ⓒ Ⓓ	22	Ⓐ Ⓑ Ⓒ Ⓓ	42	Ⓐ Ⓑ Ⓒ Ⓓ
3	Ⓐ Ⓑ Ⓒ Ⓓ	23	Ⓐ Ⓑ Ⓒ Ⓓ	43	Ⓐ Ⓑ Ⓒ Ⓓ
4	Ⓐ Ⓑ Ⓒ Ⓓ	24	Ⓐ Ⓑ Ⓒ Ⓓ	44	Ⓐ Ⓑ Ⓒ Ⓓ
5	Ⓐ Ⓑ Ⓒ Ⓓ	25	Ⓐ Ⓑ Ⓒ Ⓓ	45	Ⓐ Ⓑ Ⓒ Ⓓ
6	Ⓐ Ⓑ Ⓒ Ⓓ	26	Ⓐ Ⓑ Ⓒ Ⓓ	46	Ⓐ Ⓑ Ⓒ Ⓓ
7	Ⓐ Ⓑ Ⓒ Ⓓ	27	Ⓐ Ⓑ Ⓒ Ⓓ	47	Ⓐ Ⓑ Ⓒ Ⓓ
8	Ⓐ Ⓑ Ⓒ Ⓓ	28	Ⓐ Ⓑ Ⓒ Ⓓ	48	Ⓐ Ⓑ Ⓒ Ⓓ
9	Ⓐ Ⓑ Ⓒ Ⓓ	29	Ⓐ Ⓑ Ⓒ Ⓓ	49	Ⓐ Ⓑ Ⓒ Ⓓ
10	Ⓐ Ⓑ Ⓒ Ⓓ	30	Ⓐ Ⓑ Ⓒ Ⓓ	50	Ⓐ Ⓑ Ⓒ Ⓓ
11	Ⓐ Ⓑ Ⓒ Ⓓ	31	Ⓐ Ⓑ Ⓒ Ⓓ		
12	Ⓐ Ⓑ Ⓒ Ⓓ	32	Ⓐ Ⓑ Ⓒ Ⓓ		
13	Ⓐ Ⓑ Ⓒ Ⓓ	33	Ⓐ Ⓑ Ⓒ Ⓓ		
14	Ⓐ Ⓑ Ⓒ Ⓓ	34	Ⓐ Ⓑ Ⓒ Ⓓ		
15	Ⓐ Ⓑ Ⓒ Ⓓ	35	Ⓐ Ⓑ Ⓒ Ⓓ		
16	Ⓐ Ⓑ Ⓒ Ⓓ	36	Ⓐ Ⓑ Ⓒ Ⓓ		
17	Ⓐ Ⓑ Ⓒ Ⓓ	37	Ⓐ Ⓑ Ⓒ Ⓓ		
18	Ⓐ Ⓑ Ⓒ Ⓓ	38	Ⓐ Ⓑ Ⓒ Ⓓ		
19	Ⓐ Ⓑ Ⓒ Ⓓ	39	Ⓐ Ⓑ Ⓒ Ⓓ		
20	Ⓐ Ⓑ Ⓒ Ⓓ	40	Ⓐ Ⓑ Ⓒ Ⓓ		

ISEE Middle Level Practice Test 1

Mathematics

Quantitative Reasoning

- ❖ 37 Questions.
- ❖ Total time for this test: 35 Minutes.
- ❖ You may NOT use a calculator for this test.

Released

ISEE Middle-Level Subject Test Mathematics

1) Jim purchased a table for 40% off and saved $36. What was the original price of the table?

   A. $95

   B. $55

   C. $90

   D. 60

2) A $24 shirt now selling for $18 is discounted by what percent?

   A. 20 %

   B. 16 %

   C. 25 %

   D. 22 %

3) Which of the following shows the numbers in descending order?

   A. $\frac{1}{11}, \frac{1}{4}, \frac{1}{8}, \frac{1}{5}$

   B. $\frac{1}{8}, \frac{1}{4}, \frac{1}{5}, \frac{1}{8}$

   C. $\frac{1}{4}, \frac{1}{5}, \frac{1}{8}, \frac{1}{11}$

   D. $\frac{1}{11}, \frac{1}{5}, \frac{1}{8}, \frac{1}{4}$

4) 759,351,188 × 0.0001?

   A. 759,351.188

   B. 75,935.1188

   C. 7,593,511.88

   D. 75,935,118.8

5) If $f = x + 6y$ and $g = 4x + 2y$, what is $2f - 3g$?

   A. $12x - 4y$

   B. $10x - 6y$

   C. $-10x + 6y$

   D. $-12x + 4y$

6) Solve. $\frac{-35 \times 0.6}{6}$

   A. -2.5

   B. -3.5

   C. 6.5

   D. 6

ISEE Middle-Level Subject Test Mathematics

7) What is the value of x in the following equation? $8^x = 512$

   A. 4

   B. 2

   C. 3

   D. 5

8) The score of Emma was one eighth as that of Ava and the score of Mia was twice that of Ava. If the score of Mia was 80, what is the score of Emma?

   A. 5

   B. 14

   C. 8

   D. 15

9) The area of a circle is 49π. What is the circumference of the circle?

   A. 21π

   B. 14π

   C. 28π

   D. 7π

10) Round off the result of 3.18×7.5 to the nearest tenth?

    A. 33.9

    B. 23.9

    C. 25

    D. 35.4

11) What is the value of x in the following figure?

    A. 18°

    B. 106°

    C. 98°

    D. 108°

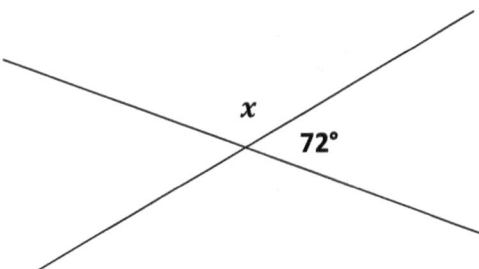

ISEE Middle-Level Subject Test Mathematics

12) What is the mean in the following set of numbers?

$$15, 28, 35, 45, 63, 95, 125$$

A. 46

B. 47

C. 52

D. 58

13) Two ninth of 54 is equal to $\frac{4}{7}$ of what number?

A. 38

B. 19

C. 42

D. 21

14) In three successive hours, a car travels 45 km, 56 km and 63 km. In the next three hours, it travels with an average speed of 52 km per hour. Find the total distance the car traveled in 6 hours.

A. 312 km

B. 328 km

C. 156 km

D. 320 km

15) The perimeter of the trapezoid below is 82. What is its area?

A. 242 cm²

B. 107.5 cm²

C. 430 cm²

D. 215 cm²

ISEE Middle-Level Subject Test Mathematics

16) Find $\frac{1}{6}$ of $\frac{3}{8}$ of 320?

   A. 35

   B. 20

   C. 15

   D. 25

17) A company pays its employee $11,000 plus 8% of all sales profit. If x is all sold profit, which of the following represents the employee's revenue?

   A. $0.08x - 11,000$

   B. $0.92x - 11,000$

   C. $0.08x + 11,000$

   D. $0.92x + 11,000$

18) Which of the following is a correct statement?

   A. $\frac{5}{8} > 0.8$

   B. $40\% = \frac{1}{4}$

   C. $3 < \frac{9}{4}$

   D. $\frac{4}{5} > 0.5$

19) What is the value of x in the following figure?

   A. 135

   B. 150

   C. 147

   D. 78

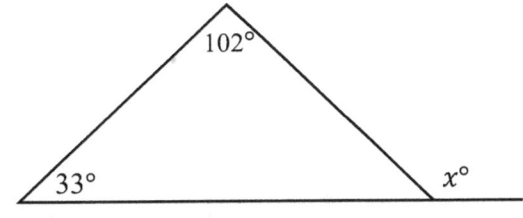

20) What is the area of a square whose diagonal is 16?

   A. 32

   B. 16

   C. 128

   D. 256

ISEE Middle-Level Subject Test Mathematics

21) The price of a laptop is decreased by 75% to $700. What is its original price?

   A. $1,400

   B. $870

   C. $770

   D. $2,800

22) The ratio of boys and girls in a class is 3:7. If there are 80 students in the class, how many more boys should be enrolled to make the ratio 1:1?

   A. 28

   B. 56

   C. 32

   D. 24

23) What is the value of x in the following equation?

   $5(x + 3) = 4(x - 2) + 15$

   A. 8

   B. -8

   C. -7

   D. 7

24) What is value of $-47 - (-61)$?

   A. 14

   B. -14

   C. -59

   D. 59

25) Car A use 4-liter petrol per 80 kilometers; car B use 3-liter petrol per 80 kilometers. If both cars drive 240 kilometers, how much more petrol does car A use?

   A. 21

   B. 9

   C. 3

   D. 6

ISEE Middle-Level Subject Test Mathematics

Quantitative Comparisons

Direction: Questions 26 to 37 are Quantitative Comparisons Questions. Using the information provided in each question, compare the quantity in column A to the quantity in Column B. Choose on your answer sheet grid.

    A if the quantity in Column A is greater.

    B if the quantity in Column B is greater.

    C if the two quantities are equal.

    D if the relationship cannot be determined from the information given.

26)

Column A	Column B
$9 + 4 \times 5 + 3$	$10 + 4 \times 6 - 9$

27) $y = -5x - 12$

Column A	Column B
The value of x when $y = 3$	-2

28)

Column A	Column B
$\sqrt{36} + \sqrt{36}$	$\sqrt{140}$

29) The average age of Joe, Michelle, and Nicole is 65.

Column A	Column B
The average age of Nicole and Michelle	The average age of Michelle and Joe

ISEE Middle-Level Subject Test Mathematics

30)

Column A	Column B
$\sqrt{285-90}$	$\sqrt{324}-\sqrt{25}$

31) A right cylinder with radius 4 inches has volume 96π cubic inches.

Column A	Column B
The height of the cylinder	3 inches

32)

Column A	Column B
$\dfrac{x^6}{6}$	$\left(\dfrac{x}{6}\right)^6$

33) x is an integer greater than zero?

Quantity A	Quantity B
$\dfrac{3}{x}+3x$	20

34) $3x^2 + 33 = 81$

$34 - 7y = 13$

Quantity A	Quantity B
x	y

35) $\dfrac{4}{5} < x < \dfrac{10}{11}$

Quantity A	Quantity B
x	$\dfrac{7}{8}$

36) The average of 10, 8, and x is 9.

Quantity A	Quantity B
x	average of $x-2, x, x+2, 3x$

37) a and b are real numbers.

$$a < b$$

Quantity A	**Quantity B**
$|a - 3b|$	$|3b - a|$

STOP

IF YOU FINISH BEFORE TIME IS CALLED, YOU MAY CHECK YOUR WORK ON THIS SECTION ONLY. DO NOT TURN TO ANY OTHER SECTION IN THE TEST.

ISEE Middle Level Practice Test 1

Mathematics

Mathematics Achievement

- ❖ **47 Questions.**
- ❖ **Total time for this test: 40 Minutes.**
- ❖ **You may NOT use a calculator for this test.**

Released Month Year

ISEE Middle-Level Subject Test Mathematics

1) $9\left(\frac{1}{6}-\frac{1}{9}\right)+7?$

   A. 9.5

   B. 16.5

   C. 7.5

   D. 11.5

2) What number is 12 more than 19% of 300?

   A. 47

   B. 69

   C. 57

   D. 79

3) If a box contains red and blue balls in ratio of 2: 7, how many red balls are there if 210 blue balls are in the box?

   A. 140

   B. 60

   C. 70

   D. 120

4) In a bundle of 80 pencils, 44 are red and the rest are blue. About what percent of the bundle is composed of blue pencils?

   A. 36%

   B. 52%

   C. 40%

   D. 45%

5) What is the value of x in the following equation?

$$(x+4)^5 = 32$$

   A. 6

   B. −6

   C. 2

   D. −2

ISEE Middle-Level Subject Test Mathematics

6) What number is 9 less than 60% of 30?

   A. 24

   B. 27

   C. 9

   D. 18

7) When a number is subtracted from 64 and the difference is divided by that number, the result is 7. What is the value of the number?

   A. 7

   B. 10

   C. 8

   D. 16

8) What is the difference in perimeter between a 7 cm by 6 cm rectangle and a circle with diameter of 6 cm? ($\pi = 3$)

   A. 8 cm

   B. 4 cm

   C. 18 cm

   D. 10 cm

9) The price of a car was $15,000 in 2014, $12,000 in 2015 and $9,600 in 2016. What is the rate of depreciation of the price of car per year?

   A. 18 %

   B. 20 %

   C. 28 %

   D. 12 %

10) 85 is equal to?

    A. $25 - (3 \times 15) + (4 \times 24)$

    B. $\left(\frac{16}{8} \times 35\right) + \left(\frac{8}{7} \times 7\right)$

    C. $\left(\left(\frac{5}{2} + 20\right) \times \frac{16}{5}\right) + 13$

    D. $(2 \times 18) + (30 \times 4) - 48$

ISEE Middle-Level Subject Test Mathematics

11) If $\frac{7x}{2} = 56$, then $\frac{7x}{4} =$?

   A. 44

   B. 14

   C. 28

   D. 16

12) Which of the following angles can represent the three angles of an isosceles right triangle?

   A. 35°, 90°, 55°

   B. 60°, 50°, 70°

   C. 45°, 45°, 90°

   D. 60°, 90°, 30°

13) In following rectangle which statement is true?

   A. The sum of all the angles equals 180°.

   B. Length of AB equal to length DC.

   C. AB is parallel to BC.

   D. AB is perpendicular to DC.

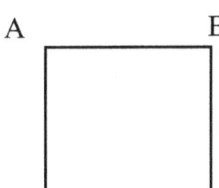

14) When a gas tank can hold 72 gallons, how many gallons does it contain when it is $\frac{5}{8}$ full?

   A. 120

   B. 72

   C. 15

   D. 45

15) Which of the following is the greatest number?

   A. $\frac{1}{4}$

   B. $\frac{4}{7}$

   C. 0.7

   D. 63%

16) A football team had $28,000 to spend on supplies. The team spent $16,000 on new balls. New sport shoes cost $110 each. Which of the following inequalities represent how many new shoes the team can purchase?

A. $110x + 16,000 \leq 28,000$

B. $110x + 16,000 \geq 28,000$

C. $16,000x + 110 \leq 28,000$

D. $16,000x + 110 \geq 28,000$

17) Calculate the approximate area of the following circle.

A. 49 cm²

B. 96.8 cm²

C. 153.86 cm²

D. 143.96 cm²

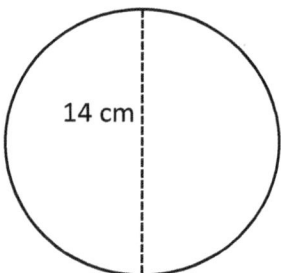

18) If 134 % of a number is 67, then what is the 60 % of that number?

A. 52

B. 35

C. 40

D. 30

19) What is the missing term in the given numbers?

10, 11, 13, 16, 20, 25, 31, ___, 46

A. 42

B. 35

C. 36

D. 38

20) The capacity of a red box is 5% greater than a blue box. If the capacity of the red box is 21 books, how many books can be put in the blue box?

A. 28 C. 20

B. 52 D. 15

21) From last year, the price of gasoline has increased from $1.14 per gallon to $2.28 per gallon. The new price is what percent of the original price?

A. 120 % C. 200 %

B. 220 % D. 210 %

22) 201 minutes = …?

A. 3.35 Hours C. 2.55 Hours

B. 3.5 Hours D. 0.4 Hours

23) If a gas tank can hold 104 gallons, how many gallons does it contain when it is $\frac{7}{8}$ full?

A. 86 C. 96

B. 71 D. 91

24) Which of the following is **not** a prime number?

A. 101 C. 17

B. 59 D. 111

25) What is the perimeter of a square that has an area of 16 square inches?

A. 256 inches C. 16 inches

B. 64 inches D. 32 inches

26) $\left(((-13)+17)\times 5\right)+(-12)$?

A. 8   C. 20

B. 12   D. 4

27) Jason left a $36.00 tip on a lunch that cost $80.00, approximately what percentage was the tip?

A. 0.36%   C. 0.45%

B. 36%   D. 45%

28) Five-kilograms apple and Seven-kilograms orange cost $29.5 If one-kilogram apple costs $1.7 how much does one-kilogram orange cost?

A. $3   C. $3.4

B. $7   D. $7.1

29) The width of a rectangle is $5x$ and its length is $9x$. The perimeter of the rectangle is 196. What is the value of x?

A. 5   C. 11

B. 7   D. 13

30) Jason is 28 miles ahead of Joe running at 5 miles per hour and Joe is running at the speed of 9 miles per hour. How long does it take Joe to catch Jason?

A. 5 hours   C. 9 hours

B. 7 hours   D. 3 hours

ISEE Middle-Level Subject Test Mathematics

31) If 40% of a class are girls, and 35% of girls play tennis, what percent of the class play tennis?

    A. 24%   C. 75 %

    B. 14%   D. 16 %

32) $[5 \times (-30) + 9] - (-6) + [8 \times 7] \div 4 = ?$

    A. 121   C. –161

    B. 161   D. –121

33) $\dfrac{1}{4} + \dfrac{\frac{-2}{3}}{\frac{4}{6}} = ?$

    A. $\dfrac{1}{4}$ C. $-\dfrac{3}{4}$

    B. $\dfrac{1}{8}$ D. $-\dfrac{1}{12}$

34) In a class, there are twice as many girls as boys. If the total number of students in the class is 78, how many girls are in the class?

    A. 25   C. 26

    B. 29   D. 28

35) At a Zoo, the ratio of lions to tigers is 9 to 4. Which of the following could NOT be the total number of lions and tigers in the zoo?

    A. 65   C. 130

    B. 91   D. 108

36) The price of a sofa is decreased by 28% to $324. What was its original price?

   A. $360

   B. $620

   C. $450

   D. $520

37) A shaft rotates 280 times in 4 seconds. How many times does it rotate in 16 seconds?

   A. 1,220

   B. 1,120

   C. 220

   D. 820

38) $\frac{4\times 22}{90}$ is closest estimate to?

   A. 1.7

   B. 1.4

   C. 1

   D. 0.57

39) Solving the equation: $4x - 37.4 = -81.4$?

   A. -11

   B. -13

   C. 11

   D. 13

40) A swimming pool holds 6,480 cubic feet of water. The swimming pool is 48 feet long and 15 feet wide. How deep is the swimming pool?

   A. 4 feet

   B. 7 feet

   C. 12 feet

   D. 9 feet

ISEE Middle-Level Subject Test Mathematics

41) A card is drawn at random from a standard 52–card deck, what is the probability that the card is of clubs? (The deck includes 13 of each suit clubs, diamonds, hearts, and spades)

A. $\frac{1}{52}$

B. $\frac{1}{4}$

C. $\frac{1}{8}$

D. $\frac{1}{13}$

42) Solve the following equation?

$$5^x = 3{,}125$$

A. 4

B. 5

C. 6

D. 3

43) $26.820 \div 0.009$?

A. 2.980

B. 29.80

C. 298.0

D. 2,980

44) What is the value of x in the following equation?

$$14 - 3(3x - 12 - 4x) = 65$$

A. 5

B. 3

C. 6

D. 4

45) Ella bought a pair of gloves for $42.79. She gave the clerk $50.00. How much change should she get back?

A. $7.71

B. $7.21

C. $8.12

D. $8.31

46) What is the area of the trapezoid?

A. 64

B. 32

C. 60

D. 128

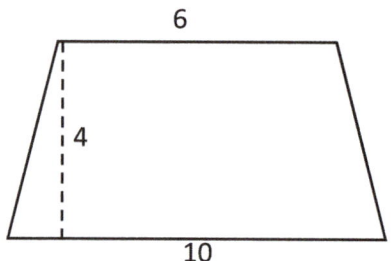

47) If 76 % of A is 19 % of B, then B is what percent of A?

A. 4%

B. 40%

C. 400%

D. 800%

ISEE Middle Level Practice Test 2

Mathematics

Quantitative Reasoning

- ❖ 37 Questions.
- ❖ Total time for this test: 35 Minutes.
- ❖ You may NOT use a calculator for this test.

Released *Month Year*

ISEE Middle-Level Subject Test Mathematics

1) What is the value of x in the following equation?

$$\frac{6^x}{36} = 1,296$$

A. 6

B. 3

C. 4

D. 9

2) Ava uses a 32% off coupon when buying a sweater that costs $29.55. If she also pays 5% sales tax on the purchase, how much does she pay?

A. 9.45

B. 9.93

C. 10.23

D. 10.87

3) An item in the store originally priced at $200 was marked down 16%. What is the final sale price of the item?

A. $168

B. $232

C. $88

D. $198

4) What's the circumference of a circle that has a diameter of 18m?

A. 56.5 m

B. 46.4 m

C. 26.5 m

D. 34.4 m

5) If the ratio of home fans to visiting fans in a crowd is 7:2 and all 36,000 seats in a stadium are filled, how many visiting fans are in attendance?

A. 18,000

B. 8,000

C. 9,000

D. 16,000

ISEE Middle-Level Subject Test Mathematics

6) In following shape y equals to?

   A. 82.5°

   B. 38.5°

   C. 47.5°

   D. 141.5°

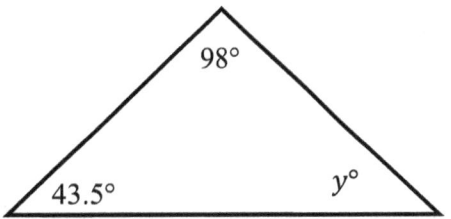

7) Which of the following shows the numbers in increasing order?

   A. $\frac{4}{5}, \frac{7}{11}, \frac{1}{6}, \frac{4}{9}$

   B. $\frac{1}{6}, \frac{4}{9}, \frac{7}{11}, \frac{4}{5}$

   C. $\frac{7}{11}, \frac{4}{9}, \frac{1}{6}, \frac{4}{5}$

   D. $\frac{7}{11}, \frac{1}{6}, \frac{4}{9}, \frac{4}{5}$

8) If an object travels at 2.5 cm per second, how many meters does it travel in 3 hours?

   A. 207

   B. 27.0

   C. 270

   D. 20.7

9) What is the area of the shaded region? (one fourth of the circle is shaded)

   Diameter = 20

   A. $100\,\pi$

   B. $50\,\pi$

   C. $10\,\pi$

   D. $25\,\pi$

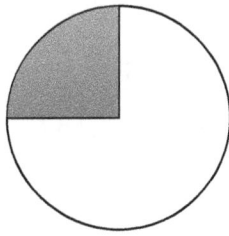

ISEE Middle-Level Subject Test Mathematics

10) Find $\frac{1}{4}$ of $\frac{1}{3}$ of $\frac{3}{7}$ of 140?

   A. 20

   B. 5

   C. 7

   D. 14

11) If a car has 70-liter petrol and after one hour driving the car use 6-liter petrol, how much petrol remaining after x-hours?

   A. $6x - 70$

   B. $70 + 6x$

   C. $70 - 6x$

   D. $6 - 70x$

12) A shirt costing $240 is discounted 25%. After a month, the shirt is discounted another 25%. Which of the following expressions can be used to find the selling price of the shirt?

   A. $(240) - (240)(1.2)$

   B. $(240) - 240(0.70)$

   C. $(240)(0.25)(0.75)$

   D. $(240)(0.75)(0.75)$

13) If $x \leq a$ is the solution of $25 + 4x \leq 53$, what is the value of a?

   A. $7x$

   B. 7

   C. -4

   D. $4x$

14) Solve for x: $12 + 3x + 8\left(\frac{x}{2}\right) = 5x + 28$

   A. 6.5

   B. 6

   C. 8

   D. 8.5

ISEE Middle-Level Subject Test Mathematics

15) The area of the trapezoid below is 210. What is the value of x?

   A. 7

   B. 11

   C. 13

   D. 23

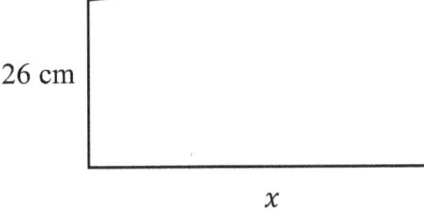

16) 9 liters of water are poured into an aquarium that's 50 cm long, 15 cm wide. How many cm will the water level in the aquarium rise due to this added water?

   (1 liter of water = 1,000 cm³)

   A. 12   C. 10

   B. 20   D. 25

17) If $4f + 4g = 8x + 4y$ and $g = 3y - 5x$, what is f?

   A. $2x+y$ C. $7x-2y$

   B. $x+7y$ D. $2y-7x$

18) In a bundle of 30 fruits, 12 are apples and the rest are bananas. What percent of the bundle is composed of apples?

   A. 20%   C. 40%

   B. 0.20%   D. 0.40%

ISEE Middle-Level Subject Test Mathematics

19) The average of 18, 22, 30 and x is 25. What is the value of x?

   A. 48

   B. 40

   C. 38

   D. 30

20) What is the perimeter of the following parallelogram?

   A. 160

   B. 188

   C. 248

   D. 98

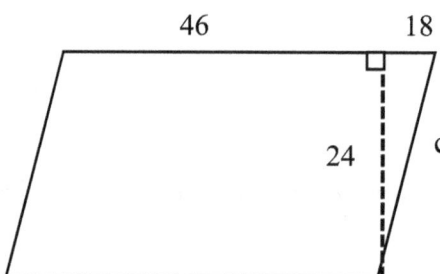

21) What is the value of $\dfrac{-\frac{7}{4} \times \frac{2}{5}}{\frac{18}{40}}$?

   A. $-\dfrac{14}{9}$

   B. $\dfrac{9}{14}$

   C. $-\dfrac{9}{14}$

   D. $-\dfrac{9}{17}$

22) What is the value of mode and median in the following set of numbers?

   $$8, 5, 9, 9, 1, 7, 6, 6, 1, 9, 1$$

   A. Mode: 1, 9 Median: 6.5

   B. Mode: 1, 9 Median: 6

   C. Mode: 9 Median: 6

   D. Mode: 1, 9 Median: 7.5

23) Which is the equivalent temperature of 95°F in Celsius?

   $$(C = \text{Celsius}) \quad C = \frac{5}{9}(F - 32)$$

   A. 35

   B. 55

   C. 20

   D. 60

24) 6 less than twice a positive integer is 42. What is the integer?

A. 30

B. 32

C. 14

D. 24

25) If Joe was making $8.15 per hour and got a raise to $8.95 per hour, what percentage increase was the raise?

A. 9.1 %

B. 0.981 %

C. 9.81 %

D. 0.098 %

ISEE Middle-Level Subject Test Mathematics

Quantitative Comparisons

Direction: Questions 26 to 37 are Quantitative Comparisons Questions. Using the information provided in each question, compare the quantity in column A to the quantity in Column B. Choose on your answer sheet grid.

A if the quantity in Column A is greater.

B if the quantity in Column B is greater.

C if the two quantities are equal.

D if the relationship cannot be determined from the information given.

26) $3x^3 - 180 = 468$

$\frac{2}{3} - \frac{y}{7} = -\frac{1}{21}$

Quantity A	Quantity B
x	y

27)

Column A	Column B
$8^2 - 2^6$	$2^6 - 8^2$

28) A computer costs $320.

Column A	Column B
A sales tax at 5% of the computer cost	$16

WWW.MathNotion.Com

ISEE Middle-Level Subject Test Mathematics

29)

Column A	Column B
$\dfrac{\sqrt{81-32}}{\sqrt{17-8}}$	$\dfrac{(9-8)}{(6-3)}$

30)

Column A	Column B
The slope of the line $8x + 4y = 28$	The slope of the line that passes through points $(3, 4)$ and $(8, -6)$

31) The sum of 4 consecutive integers is -54.

Column A	Column B
The largest of these integers	-15

32)

Column A	Column B
$\sqrt{128-33}$	$\sqrt{121} - \sqrt{49}$

33) 8 percent of x is equal to 3 percent of y, where x and y are positive numbers.

Quantity A	Quantity B
x	y

34)

Column A	Column B
The least prime factor of 30	The least prime factor of 165

35)

Quantity A	Quantity B
$(-2)^6$	2^6

36)

Quantity A	Quantity B
$(1.333)^7 (1.333)^5$	$(1.33)^{12}$

37) x is a positive number.

Quantity A	Quantity B
x^6	x^{10}

STOP

IF YOU FINISH BEFORE TIME IS CALLED, YOU MAY CHECK YOUR WORK ON THIS SECTION ONLY. DO NOT TURN TO ANY OTHER SECTION IN THE TEST.

ISEE Middle Level Practice Test 2

Mathematics

Mathematics Achievement

- ❖ 47 Questions.
- ❖ Total time for this test: 40 Minutes.
- ❖ You may NOT use a calculator for this test.

Released *Month Year*

ISEE Middle-Level Subject Test Mathematics

1) Which of the following angles is obtuse?

   A. 125 degrees

   B. 45 degrees

   C. 220 degrees

   D. 60 degrees

2) $7 + 5 \times (-4) - [12 + 18 \times 3] \div 6 = ?$

   A. 18

   B. –24

   C. –12

   D. –16

3) What is ratio of perimeter of figure A to area of figure B?

   A. $\frac{2}{7}$

   B. $\frac{2}{3}$

   C. $\frac{2}{9}$

   D. $\frac{6}{7}$

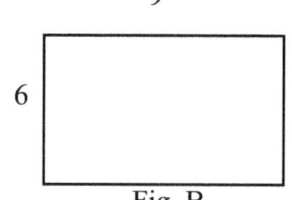
Fig. B

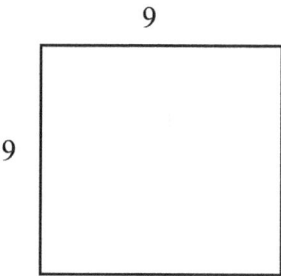
Fig. A

4) Anita's trick–or–treat bag contains 19 pieces of chocolate, 18 suckers, 14 pieces of gum, 23 pieces of licorice. If she randomly pulls a piece of candy from her bag, what is the probability of her pulling out a piece of sucker?

   A. $\frac{7}{39}$

   B. $\frac{9}{37}$

   C. $\frac{1}{17}$

   D. $\frac{1}{19}$

ISEE Middle-Level Subject Test Mathematics

5) Mr. Jones saves $3,500 out of his monthly family income of $45,500. What fractional part of his income does he save?

   A. $\frac{4}{13}$

   B. $\frac{1}{13}$

   C. $\frac{5}{26}$

   D. $\frac{3}{26}$

6) What is the difference in area between a 12 cm by 6 cm rectangle and a circle with diameter of 12 cm? ($\pi = 3$)

   A. 72

   B. 18

   C. 36

   D. 8

7) Solve the following equation?

$$(x^2 + 6x + 9) = 64$$

   A. $-11, 5$

   B. ± 8

   C. -5

   D. $8, -5$

8) 56 is equal to?

   A. $30 - (3 \times 14) + (6 \times 17)$

   B. $\left(\left(\frac{34}{8} + \frac{11}{2}\right) \times 4\right) - \frac{15}{5} + \frac{140}{7}$

   C. $\left(\frac{18}{7} \times 63\right) + (\frac{160}{8})$

   D. $(3 \times 14) + (10 \times 4.5) + 6$

ISEE Middle-Level Subject Test Mathematics

9) $\frac{12 \times 38}{8}$ is closest estimate to?

   A. 57

   B. 49

   C. 51

   D. 88

10) When a number is multiplied to itself and added by 28, the result is 53. What is the value of the number?

    A. 5 and −5

    B. 9 and −9

    C. 5

    D. 9

11) If you invest $700 at an annual rate of 6%, how much interest will you earn after one year?

    A. 4.2

    B. 4,200

    C. 420

    D. 42

12) If x = lowest common multiple of 45 and 15 then $\frac{x}{9} - 1$ equal to?

    A. 3

    B. 18

    C. 4

    D. 7

13) What is the value of x in the following equation?

    $$8^x - 188 = 324$$

    A. 6

    B. 3

    C. 9

    D. 12

ISEE Middle-Level Subject Test Mathematics

14) If angles A and B are angles of a parallelogram, what is the sum of the measures of the two angles?

   A. 90 degrees

   B. 360 degrees

   C. 180 degrees

   D. Cannot be determined.

15) Which of the following is not synonym for 16^3?

   A. 16 cubed

   B. 16 squared

   C. the square of 16

   D. 18 to the second power

16) A swing moves from one extreme point (point A) to the opposite extreme point (point B) in 25 seconds. How long does it take that the swing moves 18 times from point A to point B and returns to point A?

   A. 15 minutes

   B. 18 minutes

   C. 33 minutes

   D. 16 minutes

17) There are 2 cars moving in the same direction on a road. A red car is 27 km ahead of a blue car. If the speed of the red car is 40 km per hour and the speed of the blue car is $3\frac{1}{4}$ of the red car, how many minutes will it take the blue car to catch the red car?

   A. 24

   B. 16

   C. 18

   D. 12

ISEE Middle-Level Subject Test Mathematics

18) If the area of trapezoid is 210 what is the perimeter of the trapezoid?

   A. 60

   B. 70

   C. 56

   D. 58

   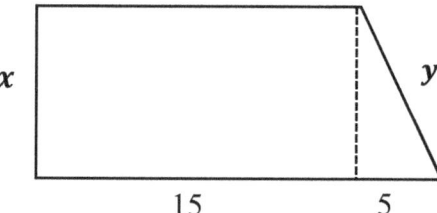

19) In two successive years, the population of a town is increased by 12% and 30%. What percent of the population is increased after two years?

   A. 44.6%   C. 42.5%

   B. 44.5%   D. 45.6%

20) In 1999, the average worker's income increased $7,000 per year starting from $38,000 annual salary. Which equation represents income greater than average?

   (I = income, x = number of years after 1999)

   A. $I > 7,000x + 38,000$ C. $I < -7,000x + 38,000$

   B. $I > -7,000x + 38,000$ D. $I < 7,000x - 38,000$

21) How many possible outfit combinations come from seven shirts, three slacks, and nine ties?

   A. 189   C. 63

   B. 168   D. 66

ISEE Middle-Level Subject Test Mathematics

22) Solving the equation: $\frac{x}{7} + \frac{3}{7} = \frac{20}{14}$?

   A. 4

   B. 7

   C. 8

   D. 12

23) What is the absolute value of the quantity four minus nine?

   A. -4

   B. 4

   C. -5

   D. 5

24) If $y = 4ab + 3b^2$, what is y when $a = 5$ and $b = 3$?

   A. 33

   B. 87

   C. 7

   D. 74

25) Which of the following angles can represent the three angles of an equilateral triangle?

   A. 60°, 60°, 60°

   B. 70°, 70°, 40°

   C. 45°, 90°, 45°

   D. 50°, 30°, 100°

26) In the following equation, what is the value of $4x - 6y$?

$$5x + 4x - 22 = 6\left(\frac{5}{6}x + y\right) - 13$$

   A. 34

   B. -34

   C. 9

   D. -9

27) Seven-kilogram apple and Nine-kilograms orange costs $92 If price of one-kilogram of apple is twice price of one-kilogram of orange. How much does one-kilogram apple cost?

A. $16

B. $8

C. $20

D. $4

28) Calculate the approximate circumference of the following circle.

A. 123.6

B. 58.8

C. 62.8

D. 98.6

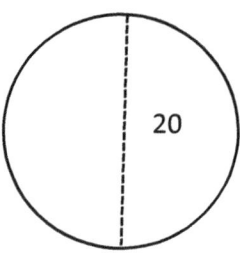

29) Which is **Not** a prime number?

A. 307

B. 97

C. 137

D. 253

30) How many tiles of 6 cm² is needed to cover a floor of dimension 8 cm by 27 cm?

A. 19

B. 32

C. 36

D. 28

31) 378 minutes=…?

A. 6.3 Hours

B. 7.2 Hours

C. 4.8 Hours

D. 7.8 Hours

32) You just drove 476 miles and it took you approximately 8 hours. How many miles per hour was your average speed?

A. about 61.5 miles per hour

B. about 59.5 miles per hour

C. about 65.5 miles per hour

D. about 57.5 miles per hour

33) In the figure below, line A is parallel to line B. What is the value of angle x?

A. 96 degree

B. 106 degree

C. 26 degree

D. 116 degree

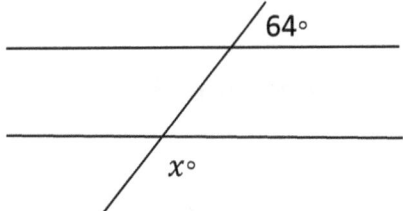

34) $\left(((-18) + 54) \times \frac{1}{4}\right) + (-14)$?

A. 8

B. 4

C. −23

D. −5

35) Three people go to a restaurant. Their bill comes to $64.00. They decided to split the cost. One person pays $ 4.20, the next person pays 5 times that amount. How much will the fourth person have to pay?

A. $40.2

B. $38.8

C. $19.8

D. $22.8

36) What is 54,773 in scientific notation?

A. 54.773×10^4

B. 45.773×10^2

C. 0.54773×10^6

D. 5.4773×10^4

37) What is the perimeter of the below right triangle?

A. 60

B. 45

C. 25

D. 65

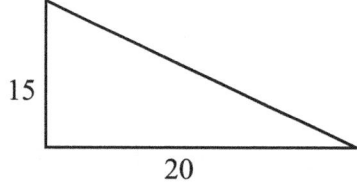

38) If 120 % of a number is equal to 36% of 80, then what is the number?

A. 24

B. 14

C. 18

D. 28

39) What is the value of $(14 - 8)!$?

A. 720

B. 6

C. 12

D. 220

40) An angle is equal to one third of its supplement. What is the measure of that angle?

A. 45

B. 236

C. 30

D. 18

ISEE Middle-Level Subject Test Mathematics

41) In a department of a company, the ratio of employees with bachelor's degree to employees with high school Diploma is 5 to 7. If there are 30 employees with bachelor's degree in this department, how many employees with High School Diploma should be moved to other departments to change the ratio of the number of employees with bachelor's degree to the number of employees with High School Diploma to 3 to 4 in this department?

A. 2

B. 42

C. 5

D. 40

42) The average weight of 19 girls in a class is 44 kg and the average weight of 21 boys in the same class is 56 kg. What is the average weight of all the 30 students in that class?

A. 49.6

B. 51.2

C. 55.5

D. 50.3

43) What is x in the following right triangle?

A. $\sqrt{12}$

B. 24

C. $\sqrt{254}$

D. $12\sqrt{5}$

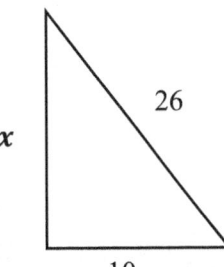

ISEE Middle-Level Subject Test Mathematics

44) John traveled 420 km in 6 hours and Alice traveled 640 km in 8 hours. What is the ratio of the average speed of John to average speed of Alice?

   A. 5: 9

   B. 7: 8

   C. 7: 9

   D. 5: 8

45) What is the difference of smallest 3–digit number and biggest 3–digit number?

   A. 991

   B. 998

   C. 990

   D. 899

46) There are three boxes, a red box, a blue box, and a yellow box. If the weight of the red box is 49 kg and the weight of the red box is 70% of the weight of the blue box, and the weight of the blue box is 140% of the weight of the yellow box, what is the weight of blue and yellow boxes respectively?

   A. 70, 50

   B. 70, 25

   C. 50, 75

   D. 50, 25

47) Each of the x students in a team may invite up to 5 friends to a party. What is the maximum number of students and guests who might attend the party?

   A. $6x + 5$

   B. $5x$

   C. $x + 6$

   D. $6x$

STOP

IF YOU FINISH BEFORE TIME IS CALLED, YOU MAY CHECK YOUR WORK ON THIS SECTION ONLY. DO NOT TURN TO ANY OTHER SECTION IN THE TEST.

Chapter 13 : Answers and Explanations

ISEE Middle Level Practice Tests

Answer Key

❋ Now, it's time to review your results to see where you went wrong and what areas you need to improve!

Practice Test 1 - Mathematics

Quantitative Reasoning						Mathematics Achievement								
1	C	16	B	31	A	1	C	16	A	31	B	46	B
2	C	17	C	32	A	2	B	17	C	32	D	47	C
3	C	18	D	33	D	3	B	18	D	33	C		
4	B	19	A	34	A	4	D	19	D	34	C		
5	C	20	C	35	D	5	D	20	C	35	D		
6	B	21	D	36	B	6	C	21	C	36	C		
7	C	22	C	37	C	7	C	22	A	37	B		
8	A	23	B			8	A	23	D	38	C		
9	B	24	A			9	B	24	D	39	A		
10	B	25	C			10	C	25	C	40	D		
11	D	26	A			11	C	26	A	41	B		
12	D	27	B			12	C	27	D	42	B		
13	D	28	A			13	B	28	A	43	D		
14	D	29	D			14	D	29	B	44	A		
15	D	30	A			15	C	30	B	45	B		

ISEE Middle Level Practice Tests

Practice Test 2 - Mathematics

Quantitative Reasoning

#	Ans	#	Ans	#	Ans
1	A	16	A	31	A
2	B	17	C	32	A
3	A	18	C	33	B
4	A	19	D	34	B
5	B	20	B	35	C
6	B	21	A	36	A
7	B	22	B	37	D
8	C	23	A		
9	D	24	D		
10	B	25	C		
11	C	26	A		
12	D	27	C		
13	B	28	C		
14	C	29	A		
15	A	30	C		

Mathematics Achievement

#	Ans	#	Ans	#	Ans	#	Ans
1	A	16	A	31	A	46	A
2	B	17	C	32	B	47	D
3	B	18	A	33	D		
4	B	19	D	34	D		
5	B	20	A	35	B		
6	C	21	A	36	D		
7	A	22	B	37	A		
8	B	23	D	38	A		
9	A	24	B	39	A		
10	A	25	A	40	A		
11	D	26	C	41	A		
12	C	27	B	42	D		
13	B	28	C	43	B		
14	D	29	D	44	B		
15	A	30	C	45	D		

ISEE Middle-Level Subject Test Mathematics

Score Your Test

ISEE scores are broken down by its four sections: Verbal Reasoning, Reading Comprehension, Quantitative Reasoning, and Mathematics Achievement. A sum of the three sections is also reported.

For the Middle Level ISEE, the score range is 760 to 940, the lowest possible score a student can earn is 760 and the highest score is 940 for each section. A student receives 1 point for every correct answer. There is no penalty for wrong or skipped questions.

The total scaled score for a Middle Level ISEE test is the sum of the scores for all sections. A student will also receive a percentile score of between 1-99% that compares that student's test scores with those of other test takers of same grade and gender from the past 3 years.

Use the next table to convert ISEE Middle level raw score to scaled score for application to 7th and 8th grade.

ISEE Middle Level Scaled Scores

Raw Score	Quantitative Reasoning		Mathematics Achievement		Raw Score	Quantitative Reasoning		Mathematics Achievement	
	7th Grade	8th Grade	7th Grade	8th Grade		7th Grade	8th Grade	7th Grade	8th Grade
0	760	760	760	760	26	900	885	885	865
1	770	765	770	765	27	905	890	885	865
2	780	770	780	770	28	910	895	890	870
3	790	775	790	775	29	910	900	890	870
4	800	780	800	780	30	915	905	895	875
5	810	785	810	785	31	920	910	895	875
6	820	790	820	790	32	925	915	900	880
7	825	795	825	795	33	930	920	900	880
8	830	800	830	800	34	930	925	905	885
9	835	805	835	805	35	935	930	905	885
10	840	810	840	810	36	935	935	910	890
11	845	815	845	815	37	940	940	910	890
12	850	820	850	820	38			915	895
13	855	825	855	825	39			920	900
14	860	830	855	830	40			925	905
15	865	835	860	835	41			925	910
16	870	840	860	840	42			930	915
17	875	845	865	840	43			930	920
18	880	845	865	845	44			935	925
19	880	850	870	845	45			935	930
20	885	855	870	850	46			940	935
21	885	860	875	850	47			940	940
22	890	865	875	855					
23	890	870	875	855					
24	895	875	880	860					
25	895	880	880	860					

ISEE Middle-Level Subject Test Mathematics

Answers and Explanations
ISEE - Middle Level
Practice Tests 1: Quantitative Reasoning

1) Answer: C.

40% off equals $36. Let x be the original price of the table. Then:

$40\% \ of \ x = 36 \rightarrow 0.4x = 36 \rightarrow x = \frac{36}{0.4} = 90$

2) Answer: C.

Use the formula for Percent of Change: $\frac{New \ Value - Old \ Value}{Old \ Value} \times 100 \ \%$

$= \frac{18-24}{24} \times 100 \ \% = -25 \ \%$

(negative sign here means that the new price is less than old price)

3) Answer: C.

$\frac{1}{5} = 0.2$ $\frac{1}{11} \cong 0.091$ $\frac{1}{4} = 0.25$ $\frac{1}{8} = 0.125$

4) Answer: B.

$759{,}351{,}188 \times \frac{1}{10{,}000} = 75{,}935.1188$

5) Answer: C.

$2f = 2 \times (x + 6y) = 2x + 12y$

$2f - 3g = 2x + 12y - 12x - 6y = -10x + 6y$

6) Answer: B.

$\frac{-35 \times 0.6}{6} = -\frac{35 \times \frac{3}{5}}{6} = -\frac{\frac{105}{5}}{6} = -\frac{105}{30} = -3.5$

7) Answer: C.

Method 1: $8 = 2^3 \rightarrow 8^x = (2^3)^x = 2^{3x}$

$512 = 2^9 \rightarrow 2^{3x} = 2^9 \rightarrow 3x = 9 \rightarrow x = 3$

Method 2: $8^x = 512$

Let's review the choices provided:

A. 4 $8^x = 512 \rightarrow 8^4 = 4{,}096$

B. 2 $8^x = 512 \rightarrow 8^2 = 64$

WWW.MathNotion.Com

C. 3 $8^x = 512 \rightarrow 8^3 = 512$

D. 5 $8^x = 512 \rightarrow 8^5 = 32,768$

Choice C is correct.

8) Answer: A.

If the score of Mia was 80, therefore the score of Ava is 40. Since the score of Emma was one eighth as that of Ava, therefore, the score of Emma is 5.

9) Answer: B.

Use the formula of areas of circles.

Area of a circle= $\pi r^2 \Rightarrow 49\pi = \pi r^2 \Rightarrow 49 = r^2 \Rightarrow r = 7$

Radius of the circle is 7. Now, use the circumference formula:

Circumference $= 2\pi r = 2\pi(7) = 14\pi$

10) Answer: B.

$3.18 = \frac{318}{100}$ and $7.5 = \frac{75}{10} \rightarrow 3.18 \times 7.5 = \frac{318}{100} \times \frac{75}{10} = \frac{23,850}{1,000} = 23.85 \cong 23.9$

11) Answer: D.

Supplementary angles sum up to 180 degrees. x and 72 degrees are supplementary angles. Then: $x = 180° - 72° = 108°$

12) Answer: D.

Mean$= \frac{15+28+35+45+63+95+125}{7} = \frac{406}{7} = 58$

13) Answer: D.

Let x be the number. Write the equation and solve for x.

$\frac{2}{9} \times 54 = \frac{4}{7} \cdot x \Rightarrow \frac{2 \times 54}{9} = \frac{4x}{7}$, use cross multiplication to solve for x.

$14 \times 54 = 4x \times 9 \Rightarrow 756 = 36x \Rightarrow x = 21$

14) Answer: D.

Add the first 3 numbers. $45 + 56 + 63 = 164$

To find the distance traveled in the next 3 hours, multiply the average by number of hours.

Distance = Average × Rate = $52 \times 3 = 156$

Add both numbers. $164 + 156 = 320$

ISEE Middle-Level Subject Test Mathematics

15) Answer: D.

The perimeter of the trapezoid is 82.

Therefore, the missing side (height) is = 82 – 29 – 24 – 19= 10

Area of the trapezoid: A = $\frac{1}{2}$ h (b1 + b2) = $\frac{1}{2}$ (10) (19 + 24) = 215

16) Answer: B.

$\frac{3}{8}$ of 320 = $\frac{3}{8}$ × 320 = 120

$\frac{1}{6}$ of 120 = $\frac{1}{6}$ × 120 = 20

17) Answer: C.

Let x be the sales profit. Then, 8% of sales profit is $0.08x$. Employee's revenue:

$0.08x + 11{,}000$

18) Answer: D.

Let's review the choices:

A. $\frac{5}{8} > 0.8$ This is not a correct statement. Because $\frac{5}{8} = 0.625$ and it's less than 0.8

B. $40\% = \frac{1}{4}$ This is not a correct statement. Because 40% = 0.4 and $\frac{1}{4} = 0.25$

C. $3 < \frac{9}{4}$ This is not a correct statement. Because $\frac{9}{4} = 2.25$ and it's less than 3

D. $\frac{4}{5} > 0.5$ This is a correct statement. $\frac{4}{5} = 0.8 \rightarrow 0.5 < \frac{4}{5}$

19) Answer: A.

$x = 33 + 102 = 135$

20) Answer: C.

The diagonal of the square is 16. Let x be the side.

Use Pythagorean Theorem: $a^2 + b^2 = c^2$

$x^2 + x^2 = 16^2 \Rightarrow 2x^2 = 16^2 \Rightarrow 2x^2 = 256$

$\Rightarrow x^2 = 128 \Rightarrow x = \sqrt{128}$

The area of the square is: $\sqrt{128} \times \sqrt{128} = 128$

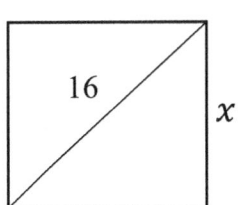

21) Answer: D.

Let x be the original price. If the price of a laptop is decreased by 75% to $700, then:

25 % of $x = 700 \Rightarrow 0.25x = 700 \Rightarrow x = 700 \div 0.25 = 2{,}800$

22) Answer: C.

The ratio of boy to girls is 3:7. Therefore, there are 3 boys out of 10 students. To find the answer, first divide the total number of students by 10, then multiply the result by 3.

$80 \div 10 = 8 \Rightarrow 8 \times 3 = 24$

There are 24 boys and 56 (80 – 24) girls. So, 32 more boys should be enrolled to make the ratio 1:1.

23) Answer: B.

Simplify: $5(x + 3) = 4(x - 2) + 15$

$5x + 15 = 4x - 8 + 15$, $5x + 15 = 4x + 7$

Subtract $4x$ from both sides: $x + 15 = 7$

Add -15 to both sides: $x + 15 - 15 = 7 - 15$, $x = -8$

24) Answer: A.

$-47 - (-61) = -47 + 61 = 61 - 47 = 14$

25) Answer: C.

Petrol of car A in 240km = $\frac{4 \times 240}{80} = 12$

Petrol of car A in 240km = $\frac{3 \times 240}{80} = 9$, $12 - 9 = 3$

26) Answer: A.

Column A: Use order of operation to calculate the result.

$9 + 4 \times 5 + 3 = 9 + 20 + 3 = 32$

Column B: $10 + 4 \times 6 - 9 \to 10 + 24 - 9 = 25$

27) Answer: B.

Column A: The value of x when $y = 3$:

$y = -5x - 12 \to 3 = -5x - 12 \to -5x = 15 \to x = -3$

Column B: -2

-2 is greater than -3.

ISEE Middle-Level Subject Test Mathematics

28) Answer: A.

Column A: Simplify.

$\sqrt{36} + \sqrt{36} = 6 + 6 = 12$

12 is greater than $\sqrt{140}$. ($\sqrt{144} = 12$)

29) Answer: D.

Column A: Based on information provided, we cannot find the average age of Nicole and Michelle or average age of Michelle and Job.

30) Answer: A.

Column A: Simplify.

$\sqrt{285 - 90} = \sqrt{195}$

Column B:

$\sqrt{324} - \sqrt{25} = 18 - 5 = 13$

$\sqrt{195}$ is bigger than 13. ($\sqrt{169} = 13$)

31) Answer: A.

Volume of a right cylinder = $\pi r^2 h \rightarrow 96\pi = \pi r^2 h = \pi(4)^2 h \rightarrow h = 6$

The height of the cylinder is 6 inches which is bigger than 3 inches.

32) Answer: A.

Simplify quantity B. **Quantity B:** $(\frac{x}{6})^6 = \frac{x^6}{6^6}$

Since the two quantities have the same numerator (x^6) and the denominator in quantity B is bigger ($6^6 > 6$), then the quantity A is greater.

33) Answer: D.

Choose different values for x and find the value of quantity A.

$x = 1$, then: Quantity A: $\frac{3}{x} + 3x = \frac{3}{1} + 3(1) = 6$

Quantity B is greater

$x = 10$, then: Quantity A: $\frac{3}{x} + 3x = \frac{3}{10} + 3(10) = 30\frac{3}{10}$

Quantity A is greater

The relationship cannot be determined from the information given.

ISEE Middle-Level Subject Test Mathematics

34) Answer: A.

$3x^2 + 33 = 81 \rightarrow 3x^2 = 81 - 33 = 48 \rightarrow x^2 = 16 \rightarrow x = \sqrt{16} = \sqrt{4^2} = 4$

$34 - 7y = 20 \rightarrow -7y = 20 - 34 = -14 \rightarrow y = \frac{-14}{-7} = 2$

35) Answer: D.

Simply change the fractions to decimals.

$\frac{4}{5} = 0.8$

$\frac{10}{11} = 0.9090 \ldots$

$\frac{7}{8} = 0.875$

As you can see, x lies between 0.8 and 0.9090… and it can be 0.86 or 0.89. The first one is less than 0.875 and the second one is greater than 0.875

The relationship cannot be determined from the information given.

36) Answer: B.

Quantity A is: $\frac{10+8+x}{3} = 9 \rightarrow x = 9$

Quantity B is: $\frac{(9-2)+9+(6+9)+(3\times 9)}{4} = 14.5$

37) Answer: C.

Choose different values for a and b and find the values of quantity A and quantity B.

$a = 4$ and $b = 5$, then:

Quantity A: $|4 - 15| = |-9| = 9$

Quantity B: $|15 - 4| = |9| = 9$

The two quantities are equal. $a = -2$ and $b = 3$, then:

Quantity A: $|-2 - 9| = |-11| = 11$

Quantity B: $|9 - (-2)| = |9 + 2| = 11$

The two quantities are equal. Any other values of a and b provide the same answer.

Answers and Explanations
ISEE - Middle Level
Practice Tests 1: Mathematics Achievement

1) Answer: C.

$9\left(\frac{1}{6}-\frac{1}{9}\right)+7=9\times\left(\frac{3-2}{18}\right)+7=\frac{9}{18}+7=\frac{1}{2}+7=\frac{15}{2}=7.5$

2) Answer: B.

19% of 300 = $\frac{19}{100}\times 300=57$

Let x be the number then, $x=57+12=69$

3) Answer: B.

$\frac{2}{7}\times 210=60$

4) Answer: D.

Number of pencils are blue = $80-44=36$

Percent of blue pencils is: $\frac{36}{80}\times 100=45\%$

5) Answer: D.

$(x+4)^5=32 \to x+4=\sqrt[5]{32}=\sqrt[5]{2^5}=2 \to x=2-4=-2$

6) Answer: C.

60% of 30 is: $\frac{60}{100}\times 30=\frac{1,800}{100}=18$

Let x be the number then: $x=18-9=9$

7) Answer: C.

Let x be the number. Write the equation and solve for x.

$(64-x)\div x=7$; Multiply both sides by x.

$(64-x)=7x$, then add x both sides. $64=8x$, now divide both sides by 8. $x=8$

8) Answer: A.

The perimeter of rectangle is: $2\times(7+6)=2\times 13=26$

The perimeter of circle is: $2\pi r=2\times 3\times\frac{6}{2}=18$

Difference in perimeter is: $26-18=8$

ISEE Middle-Level Subject Test Mathematics

9) Answer: B.

Use this formula: Percent of Change = $\frac{\text{New Value} - \text{Old Value}}{\text{Old Value}} \times 100\%$

$\frac{12,000-15,000}{15,000} \times 100\% = -20\%$ and $\frac{9,600-12,000}{12,000} \times 100\% = -20\%$

10) Answer: C.

$((\frac{5}{2} + 20) \times \frac{16}{5}) + 13 = ((\frac{5+40}{2}) \times \frac{16}{5}) + 13 = (\frac{45}{2} \times \frac{16}{5}) + 13 = 72 + 13 = 85$

11) Answer: C.

If $\frac{7x}{2} = 56$, then $7x = 112 \rightarrow x = 16$

$\frac{7x}{4} = \frac{7 \times 16}{4} = \frac{112}{4} = 28$

12) Answer: C.

All angles in a triangle sum up to 180 degrees. Then:

$2\alpha + 90° = 180° \rightarrow 2\alpha = 90 \rightarrow \alpha = 45°$

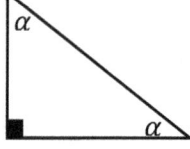

13) Answer: B.

In rectangle sides that face to face each other is equal.

14) Answer: D.

$\frac{5}{8} \times 72 = \frac{360}{8} = 45$

15) Answer: C.

$\frac{1}{4} = 0.25$ $\frac{4}{7} = 0.571$ $63\% = 0.63$

16) Answer: A.

Let x be the number of shoes the team can purchase. Therefore, the team can purchase 110 x.

The team had $28,000 and spent $16,000. Now the team can spend on new shoes $12,000 at most. Now, write the inequality: $110x + 16,000 \leq 28,000$

17) Answer: C.

Area = $\pi r^2 = \pi \times (\frac{14}{2})^2 = 49\pi = 49 \times 3.14 = 153.86$

ISEE Middle-Level Subject Test Mathematics

18) Answer: D.

First, find the number.

Let x be the number. Write the equation and solve for x.

134 % of a number is 67, then: $1.34 \times x = 67 \Rightarrow x = 67 \div 1.34 = 50$

60 % of 50 is: $0.60 \times 50 = 30$

19) Answer: D.

Find the difference of each pairs of numbers: 10, 11, 13, 16, 20, 25, 31, ___, 46

The difference of 10 and 11 is 1, 11 and 13 is 2, 13 and 16 is 3, 16 and 20 is 4, 20 and 25 is 5, 25 and 31 is 6, 31 and next number should be 7. The number is $31 + 7 = 38$

20) Answer: C.

The capacity of a red box is 5% greater than a blue box. Let x be the capacity of the blue box. Then: $x + 5\%\ of\ x = 21 \rightarrow 1.05x = 21 \rightarrow x = \frac{21}{1.05} = 20$

21) Answer: C.

The question is this: 2.28 is what percent of 1.14?

Use percent formula: Part $= \frac{percent}{100} \times$ whole

$2.28 = \frac{percent}{100} \times 1.14 \Rightarrow 2.28 = \frac{percent \times 1.14}{100}$

$\Rightarrow 228 = percent \times 1.14 \Rightarrow percent = \frac{228}{1.14} = 200$

22) Answer: A.

60 minutes = 1 Hours $\rightarrow \frac{201}{60} = 3.35$ Hours

23) Answer: D.

$\frac{7}{8} \times 104 = \frac{728}{8} = 91$

24) Answer: D.

111 is not prime number, it is divisible by 3.

25) Answer: C.

The area of the square is 16 inches. Therefore, the side of the square is square root of the area. $\sqrt{16} = 4$ inches

Four times the side of the square is the perimeter: $4 \times 4 = 16$ inches

ISEE Middle-Level Subject Test Mathematics

26) Answer: A.

$\big(((-13) + 17) \times 5\big) + (-12) = (4 \times 5) - 12 = 20 - 12 = 8$

27) Answer: D.

$36 \div 80 = 0.45 = 45\%$

28) Answer: A.

Let x be one-kilogram orange cost, then: $7x + (5 \times 1.7) = 29.5 \to 7x + 8.5 = 29.5 \to 7x = 29.5 - 8.5 \to 7x = 21 \to x = \frac{21}{7} = \$3$

29) Answer: B.

The width of a rectangle is $5x$ and its length is $9x$. Therefore, the perimeter of the rectangle is $28x$. $Perimeter\ of\ a\ rectangle = 2(width + length) = 2(5x + 9x) = 2(14x) = 28x$

The perimeter of the rectangle is 168. Then: $28x = 196 \to x = 7$

30) Answer: B.

The distance between Jason and Joe is 28 miles. Jason running at 5 miles per hour and Joe is running at the speed of 9 miles per hour. Therefore, every hour the distance is 4 miles less. $28 \div 4 = 7$

31) Answer: B.

The percent of girls playing tennis is: $40\ \% \times 35\ \% = 0.40 \times 0.35 = 0.14 = 14\ \%$

32) Answer: D.

Use PEMDAS (order of operation):

$[5 \times (-30) + 9] - (-6) + [8 \times 7] \div 4 = [-150 + 9] - (-6) + [56] \div 4 =$
$[-141] + 6 + 14 = -141 + 20 = -121$

33) Answer: C.

$\frac{1}{4} + \frac{\frac{-2}{3}}{\frac{4}{6}} = \frac{1}{4} + \frac{(-2) \times 6}{3 \times 4} = \frac{1}{4} + \frac{-12}{12} = \frac{1}{4} - 1 = \frac{1-4}{4} = -\frac{3}{4}$

34) Answer: C.

There are twice as many girls as boys. Let x be the number of girls in the class. Then:

$x + 2x = 78 \to 3x = 78 \to x = 26$

ISEE Middle-Level Subject Test Mathematics

35) Answer: D.

The ratio of lions to tigers is 9 to 4 at the zoo. Therefore, total number of lions and tigers must be divisible by 13. $9 + 4 = 13$

From the numbers provided, only 108 is not divisible by 13.

36) Answer: C.

Let x be the original price.

If the price of the sofa is decreased by 28% to \$324, then: $72\% \text{ of } x = 324 \Rightarrow$

$0.72x = 324 \Rightarrow x = 324 \div 0.72 = 450$

37) Answer: B.

Number of rotates in 16 second: $\frac{4}{280} = \frac{16}{?} \rightarrow \frac{280 \times 16}{4} = 1,120$

38) Answer: C.

$\frac{4 \times 22}{90} = \frac{88}{90} = 0.978 \cong 1$

39) Answer: A.

$4x = -81.4 + 37.4 = -44 \rightarrow x = \frac{-44}{4} = -11$

40) Answer: D.

Use formula of rectangle prism volume.

V = (length) (width) (height) $\Rightarrow 6,480 = (48)(15)(\text{height}) \Rightarrow \text{height} = 6,480 \div 720 = 9$

41) Answer: B.

The probability of choosing a club is $\frac{13}{52} = \frac{1}{4}$

42) Answer: B.

$3,125 = 5^5 \rightarrow 5^x = 5^5 \rightarrow x = 5$

43) Answer: D.

$26.820 \div 0.009 = \frac{\frac{26,820}{1,000}}{\frac{9}{1,000}} = \frac{26,820}{9} = 2,980$

44) Answer: A.

$14 - 3(3x - 12 - 4x) = 14 - 3(-x - 12) = 65 \rightarrow 14 + 3x + 36 = 65 \rightarrow 3x + 50 = 65 \rightarrow 3x = 15 \rightarrow x = 5$

45) Answer: B.

$50 - 42.79 = \$7.21$

46) Answer: B.

The area of trapezoid is: $\left(\frac{6+10}{2}\right) \times 4 = 32$

47) Answer: C.

Write the equation and solve for B:

$0.76A = 0.19B$, divide both sides by 0.19, then you will have $\frac{0.76}{0.19}A = B$, therefore:

$B = 4A$, and B is 4 times of A or it's 400% of A.

Answers and Explanations
ISEE - Middle Level
Practice Tests 2: Quantitative Reasoning

1) Answer: A.

$1,296 = 6^4 \rightarrow \frac{6^x}{36} = 6^4 \rightarrow \frac{6^x}{6^2} = 6^4 \rightarrow 6^{x-2} = 6^4 \rightarrow x - 2 = 4 \rightarrow x = 6$

2) Answer: B.

32% of $29.55 = $\frac{32}{100} \times 29.55 = \$9.456$

5% of $9.456 = $\frac{5}{100} \times 9.456 = \$0.4728$

She pays: $9.456+$0.4728 $\cong$ $9.93.

3) Answer: A.

16% of 200 = $\frac{16}{100} \times 200 = 32$

Final sale price is: $200 - 32 = \$168$

4) Answer: A.

Circumference of circle = $2\pi r = 2\pi \times \frac{18}{2} = 18\pi \cong 56.5$ m

5) Answer: B.

Number of visiting fans: $\frac{2 \times 36,000}{9} = 8,000$

6) Answer: B.

In triangle sum of all angles equal to 180° then: $y = 180° - (98° + 43.5°) = 180° - 141.5° = 38.5°$

7) Answer: B.

$\frac{1}{6} \cong 0.17$ $\frac{4}{9} \cong 0.44$ $\frac{7}{11} \cong 0.64$ $\frac{4}{5} = 0.8$

8) Answer: C.

One hour equal to 60 minutes then, 3 hours = $3 \times 60 = 180$ minutes

One minute equal to 60 seconds then, 180 minutes = $180 \times 60 = 10,800$ seconds

Distance that travels by object is: $2.5 \times 10,800 = 27,000\ cm = 270\ m$

ISEE Middle-Level Subject Test Mathematics

9) Answer: D.

Area of circle with diameter 20 is: $\pi r^2 = \pi \left(\frac{20}{2}\right)^2 = 100\pi$

The area of shaded region is: $\frac{100\pi}{4} = 25\pi$

10) Answer: B.

$\frac{3}{7}$ of 140= $\frac{3}{7} \times 140 = 60$

$\frac{1}{3}$ of 60= $\frac{1}{3} \times 60 = 20$

$\frac{1}{4}$ of 20= $\frac{1}{4} \times 20 = 5$

11) Answer: C.

The amount of petrol consumed after x hours is: $6x$

Petrol remaining: $70 - 6x$

12) Answer: D.

To find the discount, multiply the number by (100% – rate of discount).

Therefore, for the first discount we get: (240) (100% – 25%) = (240) (0.75) = 180

For the next 25 % discount: (240) (0.75) (0.75)

13) Answer: B.

$25 + 4x \leq 53 \to 4x \leq 53 - 25 \to 4x \leq 28 \to x \leq \frac{28}{4} \to x \leq 7$; Then: $a = 7$

14) Answer: C.

$12 + 3x + 8\left(\frac{x}{2}\right) = 5x + 28 \to 12 + 3x + 4x = 5x + 28 \to 2x = 16 \to x = 8$

15) Answer: A.

The area of trapezoid is: $\left(\frac{34+26}{2}\right)x = 210 \to 30x = 210 \to x = 7$

16) Answer: A.

$One\ liter = 1{,}000 cm^3 \to 9\ liters = 9{,}000\ cm^3$

$9{,}000 = 50 \times 15 \times h \to h = \frac{9{,}000}{750} = 12$ cm

17) Answer: C.

$4f + 4g = 8x + 4y \to 4f + 4(3y - 5x) = 8x + 4y \to 4f + 12y - 20x = 8x + 4y \to 4f = 8x + 4y - 12y + 20x \to 4f = 28x - 8y \to f = 7x - 2y$

ISEE Middle-Level Subject Test Mathematics

18) Answer: C.

$\frac{12}{30} \times 100 = \frac{12}{3} \times 10 = 40\%$

19) Answer: D.

Average = $\frac{\text{sum of terms}}{\text{number of terms}} \Rightarrow 25 = \frac{18+22+30+x}{4} \Rightarrow 100 = 70 + x \Rightarrow x = 30$

20) Answer: B.

$C = \sqrt{18^2 + 24^2} = \sqrt{900} = 30$

Perimeter of parallelogram = $(46 + 18 + 30) \times 2 = 188$

21) Answer: A.

$\frac{-\frac{7}{4} \times \frac{2}{5}}{\frac{18}{40}} = -\frac{\frac{7 \times 2}{20}}{\frac{18}{40}} = -\frac{\frac{14}{20}}{\frac{18}{40}} = -\frac{14}{20} \div \frac{18}{40} = -\frac{14 \times 40}{18 \times 20} = -\frac{28}{18} = -\frac{14}{9}$

22) Answer: B.

We write the numbers in the order: 1, 1, 1, 5, 6, 6, 7, 8, 9, 9, 9.

The mode of numbers is: 1 and 9; median is: 6.

23) Answer: A.

Plug in 95 for F and then solve for C.

$C = \frac{5}{9}(F - 32) \Rightarrow C = \frac{5}{9}(95 - 32) \Rightarrow C = \frac{5}{9}(63) = 35$

24) Answer: D.

Let x be the integer. Then: $2x - 6 = 42$

Add 6 both sides: $2x = 48$; Divide both sides by 2: $x = 24$

25) Answer: C.

$8.95 - 8.15 = 0.8$, $\frac{0.8}{8.15} \times 100 = 9.81$

26) Answer: A.

$3x^3 - 180 = 468 \rightarrow 3x^3 = 180 + 468 = 648 \rightarrow x^3 = \frac{648}{3} = 216 \rightarrow x =$

$\sqrt[3]{216} = \sqrt[3]{6^3} = 6$

$21 \times \left(\frac{2}{3} - \frac{y}{7} = -\frac{1}{21}\right) \Rightarrow 14 - 3y = -1 \Rightarrow -3y = -1 - 14 \Rightarrow y = 5$

ISEE Middle-Level Subject Test Mathematics

27) Answer: C.

Column A: $8^2 - 2^6 = 64 - 64 = 0$

Column B: $2^6 - 8^2 = 64 - 64 = 0$

28) Answer: C.

Column A: 5% of the computer cost is 16: $5\% \times 320 = 0.05 \times 320 = 16$

Column B: 16

29) Answer: A.

Column A: Simplify. $\frac{\sqrt{81-32}}{\sqrt{17-8}} = \frac{\sqrt{49}}{\sqrt{9}} = \frac{7}{3}$

Column B: $\frac{(9-8)}{(6-3)} = \frac{2}{3}$

30) Answer: C.

Column A: The slope of the line $8x + 4y = 28$ is -2.

Write the equation in slope intercept form.

$8x + 4y = 28 \rightarrow 4y = -8x + 28 \rightarrow y = -2x + 7$

Column B: The slope of the line that passes through points $(3, 4)$ and $(8, -6)$: Use slope formula:

$slope\ of\ a\ line = \frac{y_2 - y_1}{x_2 - x_1} = \frac{-6-4}{8-3} = -2$

31) Answer: A.

Column A: First, find the integers. Let x be the smallest integer. Then the integers are $x, (x+1), (x+2),$ and $(x+3)$. The sum of the integers is -54. Then:

$x + x + 1 + x + 2 + x + 3 = -54 \rightarrow 4x + 6 = -54 \rightarrow 4x = -60 \rightarrow x = -15$

Column B: -15

The smallest integer is -15, therefore, the largest integer is bigger than that.

32) Answer: A.

Column A: Simplify. $\sqrt{128 - 33} = \sqrt{95}$

Column B: $\sqrt{121} - \sqrt{49} = 11 - 7 = 5$

$\sqrt{95}$ is bigger than 5. ($\sqrt{25} = 5$)

ISEE Middle-Level Subject Test Mathematics

33) Answer: B.

8% of x = 3% of y → $0.08\,x = 0.03\,y$ → $x = \frac{0.03}{0.08}y$ → $x = \frac{3}{8}y$, therefore, y is bigger than x.

34) Answer: B.

prime factoring of 30 is: $2 \times 3 \times 5$

prime factoring of 165 is: $3 \times 5 \times 11$

Quantity A = 2 and Quantity B = 3

35) Answer: C.

Simplify both quantities.

Quantity A: $(-2)^6 = (-2) \times (-2) \times (-2) \times (-2) \times (-2) \times (-2) = 64$

Quantity B: $2 \times 2 \times 2 \times 2 \times 2 \times 2 = 64$

The two quantities are equal.

36) Answer: A.

Use exponent "product rule": $x^n \times x^m = x^{n+m}$

Quantity A: $(1.333)^7 (1.333)^5 = (1.333)^{7+5} = (1.333)^{12}$

Quantity B: $(1.33)^{12}$

37) Answer: D.

Choose different values for x and find the value of quantity A and quantity B.

$x = 1$, then:

Quantity A: $x^6 = 1^6 = 1$

Quantity B: $x^{10} = 1^{10} = 1$

The two quantities are equal.

$x = 2$, then: Quantity A: $x^6 = 2^6$

Quantity B: $x^{10} = 2^{10}$

Quantity B is greater.

Therefore, the relationship cannot be determined from the information given.

Answers and Explanations
ISEE - Middle Level
Practice Tests 2: Mathematics Achievement

1) Answer: A.

Angle between 90° and 180° is called obtuse angle.

2) Answer: B.

Use PEMDAS (order of operation):

$7 + 5 \times (-4) - [12 + 18 \times 3] \div 6 = 7 + 5 \times (-4) - [12 + 54] \div 6 = 7 - 20 - 66 \div 6 = -13 - 11 = -24$

3) Answer: B.

Perimeter A= $9 \times 4 = 36$

Area B= $6 \times 9 = 54 \Rightarrow \frac{36}{54} = \frac{2}{3}$

4) Answer: B.

Probability $= \frac{number\ of\ desired\ outcomes}{number\ of\ total\ outcomes} = \frac{18}{19+18+14+23} = \frac{18}{74} = \frac{9}{37}$

5) Answer: B.

3,500 out of 45,500 equals to $\frac{3,500}{45,500} = \frac{35}{455} = \frac{5}{65} = \frac{1}{13}$

6) Answer: C.

The area of rectangle is: $12 \times 6 = 72 cm^2$

The area of circle is: $\pi r^2 = \pi \times (\frac{12}{2})^2 = 3 \times 36 = 108$

Difference in area is: $108 - 72 = 36$

7) Answer: A.

$x^2 + 6x + 9 = (x+3)^2 \rightarrow (x+3)^2 = 64 \rightarrow x+3 = \pm 8$

$\rightarrow x + 3 = 8 \rightarrow x = 5$

or $x + 3 = -8 \rightarrow x = -11$

8) Answer: B.

$$\left(\left(\frac{34}{8}+\frac{11}{2}\right)\times 4\right)-\frac{15}{5}+\frac{140}{7}=\left(\left(\frac{34+44}{8}\right)\times 4\right)-\frac{15}{5}+\frac{140}{7}=\left(\left(\frac{78}{8}\right)\times 4\right)-3+$$

$+20 = (39) - 3 + 20 = 39 + 17 = 56$

9) Answer: A.

$\frac{12\times 38}{8} = \frac{456}{8} = 57$

10) Answer: A.

Let x be the number, then:

$x^2 + 28 = 53 \to x^2 = 25 \to x^2 - 25 = 0 \to (x+5)(x-5) = 0 \to x = 5 \text{ or } x = -5$

11) Answer: D.

6% of $700 = $\frac{6}{100} \times 700 = \$42$

12) Answer: C.

Prime factorizing of $45 = 3 \times 3 \times 5$

Prime factorizing of $15 = 3 \times 5$

$x = \text{LCM} = 3 \times 3 \times 5 = 45$

$\frac{45}{9} - 1 = 5 - 1 = 4$

13) Answer: B.

$8^x - 188 = 324 \to 8^x = 324 + 188 = 512$ and $512 = 8^3$

$8^x = 8^3 \to x = 3$

14) Answer: D.

All angles in a parallelogram sum up to 360 degrees. Since, we only have 2 angles, therefore the answer cannot be determined.

15) Answer: A.

16 cubed is: $16^3 = 4,096$

16) Answer: A.

Swing moves once from point A to point B and returns to point A is: 25 +25=50 seconds

ISEE Middle-Level Subject Test Mathematics

Therefore, for 18 times: $18 \times 50 = 900$ seconds

$900 \div 60 = 15$ minutes

17) Answer: C.

$3\frac{1}{4} = \frac{13}{4} = 3.25$

Speed of the blue car: $3.25 \times 40 = 130$

Difference of the cars' speed: $130 - 40 = 90$

The red car is 27 km ahead of a blue car. Therefore, it takes 18 minutes to catch the red car. $\frac{27}{90} = 0.3$ Hour= 18 minutes

18) Answer: A.

The area of trapezoid is: $\left(\frac{20+15}{2}\right) \times x = 210 \to 35x = 420 \to x = 12$

$Y = \sqrt{5^2 + 12^2} = 13$

Perimeter is: $12 + 20 + 13 + 15 = 60$

19) Answer: D.

The population is increased by 12% and 30%. 12% increase changes the population to 112% of original population. For the second increase, multiply the result by 130%.

$(1.12) \times (1.30) = 1.456 = 145.6\%$

45.6 percent of the population is increased after two years.

20) Answer: A.

Let x be the number of years. Therefore, $7,000 per year equals $7000x$.

starting from $38,000 annual salary means you should add that amount to $7,000x$.

Income more than that is: $I > 7,000\,x + 38,000$

21) Answer: A.

To find the number of possible outfit combinations, multiply number of options for each factor: $7 \times 3 \times 9 = 189$

22) Answer: B.

$\frac{x}{7} + \frac{3}{7} = \frac{20}{14} \to \frac{x}{7} = \frac{20}{14} - \frac{3}{7} = \frac{20-6}{14} = \frac{14}{14} \to \frac{x}{7} = 1 \to x = 1 \times 7 = 7$

ISEE Middle-Level Subject Test Mathematics

23) Answer: D.

$|4 - 9| = |-5| = 5$

24) Answer: B.

$y = 4ab + 3b^2$

Plug in the values of a and b in the equation: $a = 5$ and $b = 3$

$y = 4(5)(3) + 3(3)^2 = 60 + 3(9) = 60 + 27 = 87$

25) Answer: A.

The angles of an equilateral triangle are 60, 60, 60 degrees.

26) Answer: C.

$5x + 4x - 22 = 6\left(\frac{5}{6}x + y\right) - 13 \to 9x - 22 = \left(6 \times \frac{5}{6}x\right) + 6y - 13 \to 9x - 22 = 5x + 6y - 13$

$\to 9x - 5x - 6y = 22 - 13 \to 4x - 6y = 9$

27) Answer: B.

Let x be price of one-kilogram of apple and y be price of one-kilogram of orange, then: $x = 2y$

$7x + 9y = 92 \to 7(2y) + 9y = 92 \to 23y = 92 \to y = \frac{92}{23} = 4 \to x = 2 \times 4 = 8$

28) Answer: C.

Perimeter $= 2\pi r = 2 \times \pi \times \frac{20}{2} = 20\pi = 62.8$

29) Answer: D.

253 is not prime number, it is divided by 11.

30) Answer: C.

The area of the floor is: 8 cm × 27 cm = 216 cm

The number is tiles needed = 216 ÷ 6 = 36

31) Answer: A.

60 minutes = 1 Hours $\to \frac{378}{60} = 6.3$ Hours

32) Answer: B.

Average speed: $\frac{476}{8} = 59.5$ miles per hour

ISEE Middle-Level Subject Test Mathematics

33) Answer: D.

$180° - 64° = 116°$

34) Answer: D.

$\left(\left((-18) + 54\right) \times \frac{1}{4}\right) + (-14) = \left((36) \times \frac{1}{4}\right) - 14 = 9 - 14 = -5$

35) Answer: B.

Let x be the price that third person has to pay then; $64 = 4.20 + (5 \times 4.20) + x \rightarrow x = 64 - 25.2 = 38.8$

36) Answer: D.

$54{,}773 = 5.4773 \times 10^4$

37) Answer: A.

$c = \sqrt{15^2 + 20^2} = \sqrt{625} = 25$

Perimeter: $15 + 20 + 25 = 60$

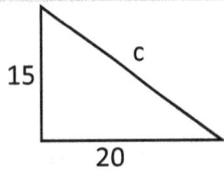

38) Answer: A.

Let x be the number then, 120 % of $x = 1.2x$

$1.2x = 0.36 \times 80 = 28.8 \rightarrow x = \frac{28.8}{1.2} = 24$

39) Answer: A.

$(14 - 8)! = 6! = 6 \times 5 \times 4 \times 3 \times 2 \times 1 = 720$

40) Answer: A.

The sum of supplement angles is 180. Let x be that angle. Therefore, $x + 3x = 180$

$4x = 180$, divide both sides by 4: $x = 45$

41) Answer: A.

Number of employees if ratio is 5 to 7: $\frac{12 \times 30}{5} = 72$

Number of employees with high school Diploma if ratio is 5 to 7: $72 - 30 = 42$

Number of employees if ratio is 3 to 4: $\frac{7 \times 30}{3} = 70$

Number of employees with high school Diploma if ratio is 3 to 4: $70 - 30 = 40$

Number of employees with High School Diploma should be moved to other departments: $42 - 40 = 2$

42) Answer: D.

$$\text{average} = \frac{\text{sum of terms}}{\text{number of terms}}$$

The sum of the weight of all girls is: $19 \times 44 = 836$ kg

The sum of the weight of all boys is: $21 \times 56 = 1,176$ kg

The sum of the weight of all students is: $836 + 1,176 = 2,012$ kg

Average $= \frac{2,012}{40} = 50.3$

43) Answer: B.

$x^2 + 10^2 = 26^2 \to x^2 = 26^2 - 10^2 \to x = \sqrt{26^2 - 10^2} \to x = \sqrt{676 - 100} = \sqrt{576} = 24$

44) Answer: B.

The average speed of john is: $420 \div 6 = 70$

The average speed of Alice is: $640 \div 8 = 80$

Write the ratio and simplify. $70 : 80 \Rightarrow 7 : 8$

45) Answer: D.

Smallest 3–digit number is 100, and biggest 3–digit number is 999. The difference is: 899.

46) Answer: A.

Weight of red $= 70\%$ Weight of blue

$49 = 0.70\, x \Longrightarrow x = \frac{49}{0.70} = 70$

Weight of blue box $= 140\%$ Weight of yellow box

$70 = 1.40 \times x \Longrightarrow x = \frac{70}{1.40} = 50$

47) Answer: D.

Since each of the x students in a team may invite up to 5 friends, the maximum number of people in the party is 6 times x or $6x$. (one student + 5 friends = 6 people)

"End"

www.ingramcontent.com/pod-product-compliance
Lightning Source LLC
Chambersburg PA
CBHW080438110426
42743CB00016B/3199